DETAILED INTERPRETATION OF CONTEMPORARY ARCHITECTURAL DESIGN

2

当代建筑设计详解

佳图文化 主编

中国林业出版社

图书在版编目（CIP）数据

当代建筑设计详解 . 2 / 佳图文化主编 . -- 北京 : 中国林业出版社 , 2018.10

ISBN 978-7-5038-9792-4

Ⅰ . ①当… Ⅱ . ①佳… Ⅲ . ①建筑设计－作品集－中国－现代 Ⅳ . ① TU206

中国版本图书馆 CIP 数据核字 (2018) 第 239858 号

中国林业出版社
责任编辑：李 顺 薛瑞琦
出版咨询： （010）83143569

出 版：中国林业出版社（100009 北京西城区德内大街刘海胡同 7 号）
网 站：http://lycb.forestry.gov.cn/
印 刷：固安县京平诚乾印刷有限公司
发 行：中国林业出版社
电 话：（010）83143500
版 次：2019 年 9 月第 1 版
印 次：2019 年 9 月第 1 次
开 本：889mm×1194mm 1 ／ 16
印 张：20
字 数：200 千字
定 价：298.00 元

Preface 前言

"Modern Chinese Architecture" series is a set of professional books that introduce and present modern Chinese architecture comprehensibly. It selects the excellent works with different design concepts and of different types that could reflect the level and development trend of contemporary Chinese architecture the most to represent the style and features. It revolves around the architectural works and analyzes further into overall planning, design concept, challenge, architectural layout, detailed design, relationship between the old and new in reconstruction, relationship between the building and the city, etc. thus to interpret the design essence from various angles. The selected works include office buildings, education buildings, commercial buildings, cultural buildings, hotel buildings, sports buildings, transportation buildings, healthcare buildings and so on. All the works are introduced with plans, elevations, sections, construction drawings, joint details and photos, presenting diverse architectural forms and details in a comprehensive and visual way for dear readers.

《当代中国建筑》系列图书为一套全面介绍和展示当代中国建筑的专业建筑类书籍。全书精选最能反映当代中国建筑水平和发展趋势的优秀作品，通过对这些不同类型、不同设计理念的建筑作品的展示和解读，反映当代中国建筑的风貌。全书内容围绕建筑设计案例展开，深入分析建筑案例的整体规划、设计构思、设计难点、建筑布局、细部设计、改扩建中新与老的联系、建筑与城市的关系等众多方面的问题，以期从各个角度展现建筑案例的设计精髓。所选案例囊括了办公建筑、教育建筑、商业建筑、文化建筑、酒店建筑、体育建筑、交通建筑、医疗建筑等众多建筑类型。案例中展示的平面图、立面图、剖面图、施工图、节点详图，以及建成实景图等，将不同的建筑形式以及建筑细节内容更为详尽、直观地呈现出来，以供关注当代中国建筑状况的读者借鉴、参考。

Contents 目录

Educational Building 教育建筑

- 010 Ambow (Dalian) Service Outsourcing Talent Training Base
  安博（大连）服务外包人才实训基地

- 014 Dalian Harbour Affairs College's Jin Shi Tan Campus
  大连枫叶职业技术学院金石滩校区

- 020 Dalian University of Technology Chemical Complex Building
  大连理工大学化工综合楼

- 026 Erdos No.1 Middle School Information Center
  鄂尔多斯市第一中学图文信息中心

- 034 Hengdan School in Wenxian, Gansu
  甘肃文县横丹学校

- 038 Jiang Nan Shui Du Middle School
  江南水都中学

- 044 New Campus of Jiangyin High School
  江苏省江阴高级中学新校区

- 050 The North No.2 Liberal Arts Building, Nanjing University of Aeronautics and Astronautics (NUAA) Jiangjun Road East Campus　南航将军路校区东区北 2 号文科楼

- 058 Century Hall of Tsinghua University
  清华百年会堂

- 064 Academy for International Business Officials (AIBO), MOFCOM
  商务部国际商务官员研修学院

072	New Campus of Shenzhen Institute of Information and Technology 深圳信息职业技术学院迁址新建工程
078	Weifang Municipal Party Committee Party School Teaching & Administration Department 潍坊市市委党校教学行政组团
084	Wuxi City College of Vocational Technology New Campus 无锡城市职业技术学院新校区
090	Central South University Comprehensive Teaching Building 中南大学综合教学楼
096	International Students Apartment of Peking University 北京大学留学生公寓
102	New Library of Nanjing Arts Institute 南京艺术学院新图书馆
106	Chengde Petroleum Higher Academy Library 承德高等石油专科学院图书馆
112	Dushuhu Higher Education District Southeast University Suzhou Graduate School Phase II 独墅湖高教区东南大学苏州研究生院二期工程
122	Arts Centre of Shandong University of Technology 山东理工大学大学生艺术中心

Contents 目录

- 136   Library of Shandong University of Technology
  山东理工大学图书馆

- 142   Fujian South China Women's Vocational College New Campus
  福建华南女子职业学院大学城新校区

- 148   Yifu Library, West Lake Campus of Fuyang Normal College
  阜阳师范学院西湖校区逸夫图书馆综合楼

- 154   Sichuan Branch of the National Prosecutors College
  国家检察官学院四川分院

- 162   Liaoning College of Communications Library
  辽宁交通高等专科学校图书馆

- 170   Library of Shandong Union Management Cadres College
  山东省工会管理干部学院图书馆

- 178   Library of Changchun University of Chinese Medicine
  长春中医药大学图书馆

- 184   North Campus of Zhoukou Normal University Library Management Center
  周口师范学院北校区图书馆管理中心

Transportation Building 交通建筑

- 192   Daqing Highway Passenger Terminal
  大庆市公路客运枢纽站

- 198   Terminal of Jieyang Chaoshan International Airport
  揭阳潮汕机场航站楼及配套工程

204	Suzhou Automobile North Station Reconstruction Project 苏州汽车北站改建工程
212	Suzhou Rail Transit Line 1 Control Center 苏州市轨道交通一号线工程控制中心大楼
218	Turpan Jiaohe Airport Terminal 吐鲁番机场航站楼工程

Healthcare Building 医疗建筑

224	Beijing Chaoyang Hospital Reconstruction & Extension 北京朝阳医院改扩建工程门急诊及病房楼
230	Shenyang Air Force 1221 Project 沈空1221工程（文体医疗综合楼）
236	Chongzhou People's Hospital and Chongzhou Maternal and Child Health Hospital 崇州市人民医院及崇州市妇幼保健院
242	Outpatient Building of Jinan Military General Hospital 济南军区总医院门诊综合楼
246	Lingnan Hospital of the Third Hospital Affiliated Hospital, Sun Yat-Sen University (Luogang Central Hospital) 中山大学附属第三医院岭南医院（萝岗中心医院）
250	Pengzhou People's Hospital 彭州市人民医院

Educational Building
教育建筑

- Human Scale
  尺度适宜

- Multiple Functions
  功能齐全

- Safe and Efficient
  高效安全

- Green and Environment-friendly
  绿色环保

KEY WORDS 关键词

Fully Functional 功能完备
Succinct Facade 立面简洁
Humanization 人性化

Ambow (Dalian) Service Outsourcing Talent Training Base
安博（大连）服务外包人才实训基地

FEATURES 项目亮点

This single building was designed to meet the function better, to provide a comfortable place for living and a vital "social center" for exchange and communication. The façade highlights the features of IT industry such as succinctness, emphasizing regulation, modernity and humanization.

项目单体建筑设计以更好地满足使用功能为出发点，立面设计突出IT行业简洁的特点，强调条理性、富有时代感、人性化等特点，项目力求创造一个舒适的生活场所，一个充满活力且能促进交流的"社交中心"。

Location: Dalian, Liaoning, China
Architectural Design: Dalian Architectural Design & Research Institute Co., Ltd.
Total Floor Area: 117,543.71 m²

项目地点：中国辽宁省大连市
建筑设计：大连市建筑设计研究院有限公司
总建筑面积：117 543.71 m²

Overview

This group of education buildings accommodate 12 single multi-storey buildings including teaching building, canteen, auditorium and dormitory, etc. which cover a land of 73,100 m² and boast a plot ratio of 1.56. It aims to create comfortable place for living and a vital "social center" for exchange and communication.

项目概况

该项目为教育建筑群，包括教学楼、食堂、报告厅、宿舍等共计12栋单体多层建筑。该项目用地面积为73 100 m²，容积率为1.56。项目力求创造一个舒适的生活场所，一个充满活力且能促进交流的"社交中心"。

Single Building

The single building was designed to meet the function better, to provide users the greatest resource at low cost (2,400 yuan for per square meter, decoration included). And the facade highlights the features of IT industry such as succinctness, emphasizing regulation, modernity and humanization.

单体建筑

单体建筑平面设计以如何更好地满足使用功能为出发点，在低造价（每平方米2 400元，含装修）的大前提下，为使用者提供最丰富的资源。单体建筑立面设计突出IT行业简洁的特点，强调条理性、富有时代感、人性化等特点。

Site Plan 总平面图

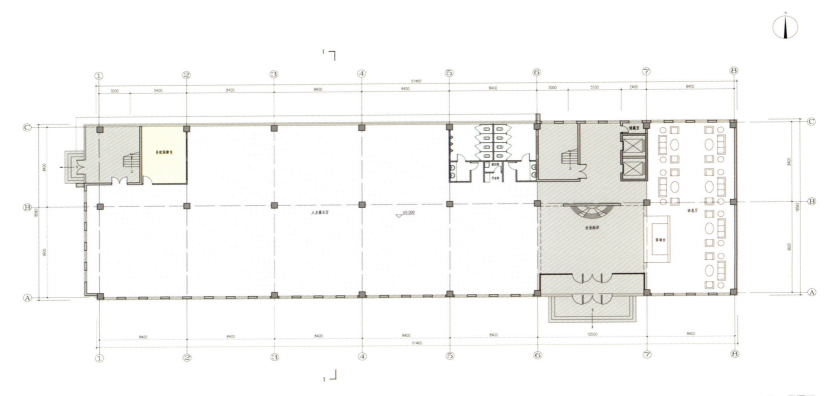

Plan 平面图

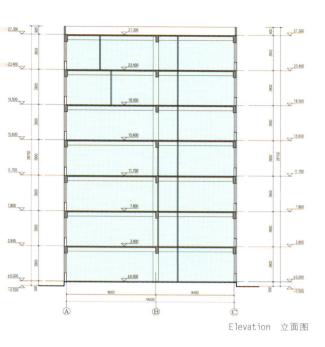

Elevation 立面图

KEY WORDS 关键词

Completed Functions 功能齐全
Well-planned 规划良好
Appealing Design 造型优美

Dalian Harbour Affairs College's Jin Shi Tan Campus
大连枫叶职业技术学院金石滩校区

FEATURES 项目亮点

This plot is divided into two parts, in the middle of which is an urban road. The southwestern side is adjacent to the main passage of Jin Shi Tan — Central Street, and the eastern side is mountains with good natural vegetation.

项目用地分为东西两个地块，之间以一条城市道路为分界，西南面毗邻金石滩对外联系的主要通道——中心大街，东面为自然山体，植被覆盖良好。

Location: Dalian City, Liaoning Province, China
Architectural Design: Dalian Architectural Design and Research Institute Co.,Ltd.

项目地点：中国辽宁省大连市
设计单位：大连市建筑设计研究院有限公司

Overview

This project locates in Jin Shi Tan, Dalian city, the shore of Yellow Sea, and is adjacent to Jin Shi Tan Geological Park with wonderful geographical environment. It is a Canada-China cooperation project, a boarding polytechnic institute. The campus is planned to hold 10,000 people in total, among which are 9,000 students and 1,000 faculty members.

项目概况

大连枫叶职业技术学院校区规划项目位于大连市金石滩，地处黄海之滨，毗邻金石滩地质公园，地理环境优美。该项目为中加合作项目，寄宿式高等职业技术学院。校区计划容纳总人数为10 000人，其中学生9 000人，教职工1 000人。

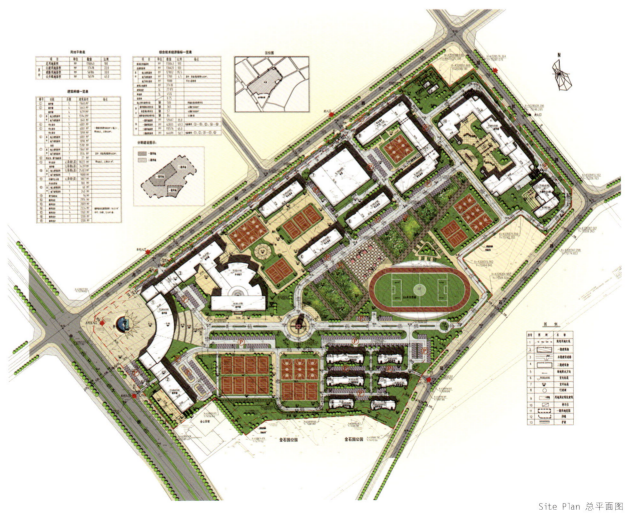

Site Plan 总平面图

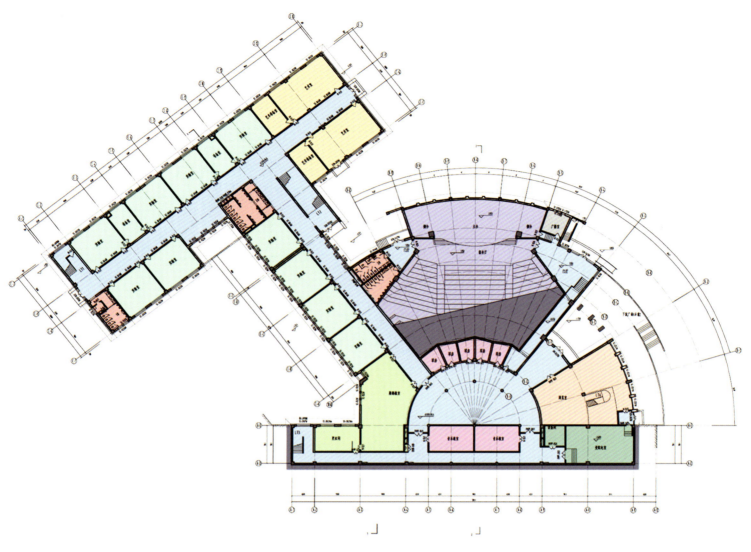

First Floor Plan 一层平面图

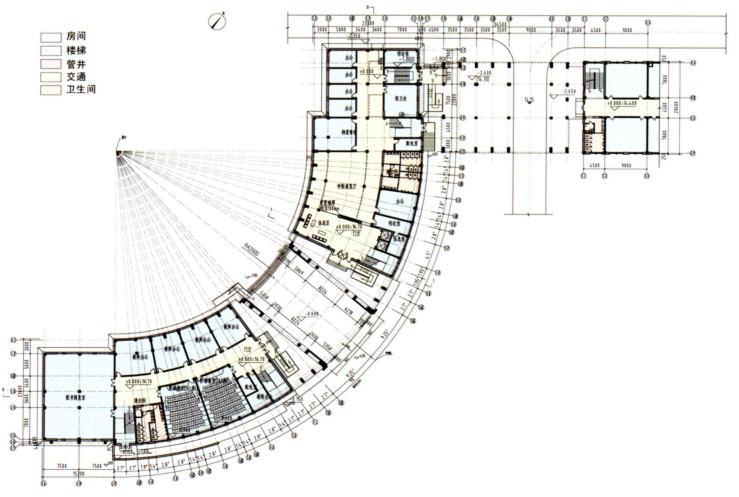

First Floor Plan of NO.1 Building 1号楼一层平面图

Planning and Layout

This plot is divided into two parts, in the middle of which is an urban road. The southwestern side is adjacent to the main passage of Jin Shi Tan — Central Street, and the eastern side is mountains with good natural vegetation.

The total land area is about 400,000 m², and will be constructed in two phases, in one of which the land area will be 170,000 m² and the floor area will be 12,700 m². There will be main teaching building, complex teaching building, student dormitory building, teacher dormitory building and the complex service building equipped with canteen, indoor stadium etc.

规划布局

项目用地分为东西两个地块,之间以一条城市道路为分界,西南面毗邻金石滩对外联系的主要通道——中心大街,东面为自然山体,植被覆盖良好。规划总用地面积约为 400 000 m²,分两期建设。其中一期工程用地面积约 170 000 m²,建筑面积 12 700 m²。设有主教学楼、综合教学楼、学生宿舍楼、教师宿舍楼、带有食堂和室内运动场等设施的综合服务楼等单体项目。

Sectional Drawing 剖面图

Sectional Drawing 剖面图

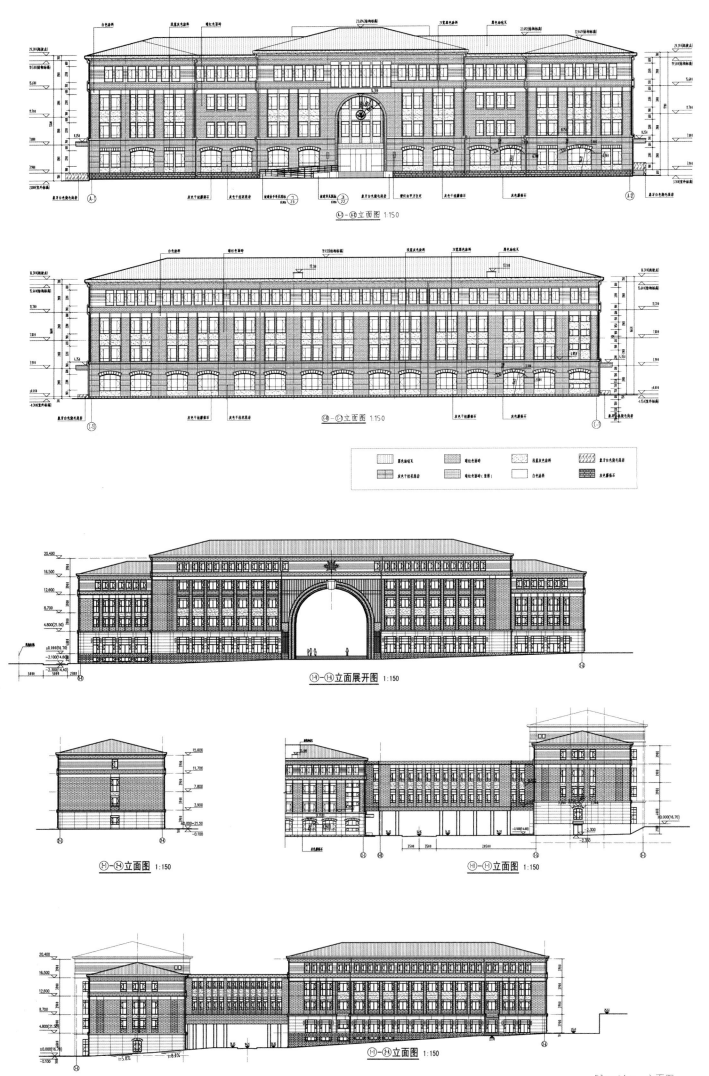

Elevation 立面图

KEY WORDS 关键词

Rhythmic Beauty 节奏美感

Vivid Color 色彩鲜明

Square Volume 体量方正

Dalian University of Technology Chemical Complex Building
大连理工大学化工综合楼

FEATURES 项目亮点

Rational rhythm of unified column width reflects a kind of rhythmic beauty and forms the cultural traits of this campus building.

建筑统一的柱网开间产生理性的韵律，体现了一种节奏美感，形成校园建筑的文化特质。

Location: Dalian, Liaoning, China
Architectural Design: Architectural Design and Research Institute of Tsinghua University Co., Ltd.
Land Area: 18,600 m²
Monomer Construction Area: 16,390 m²
Height: 27.5 m
Plot Ratio: 0.88

项目地点：中国辽宁省大连市
设计单位：清华大学建筑设计研究院有限公司
用地面积：18 600 m²
单体建筑面积：16 390 m²
建筑高度：27.5 m
容积率：0.88

Overview

This building is located in the middle of the new land to the west of the headquarters of Dalian University of Technology, mainly for teaching, research, experiment and office space. The capital resource is comprised of four parts: central government funding, "985 Project" Phase II Liaoning Province supporting fund, "985 Project" Phase II Dalian City supporting fund and school self-financing.

项目概况

大连理工大学化工综合楼位于凌水校区本部西侧新征土地的中部，项目主要用途为教学、科研、实验及办公用房。资金来源由中央财政拨款、"985工程"二期辽宁省配套经费、"985工程"二期大连市配套经费和学校自筹经费等四部分构成。

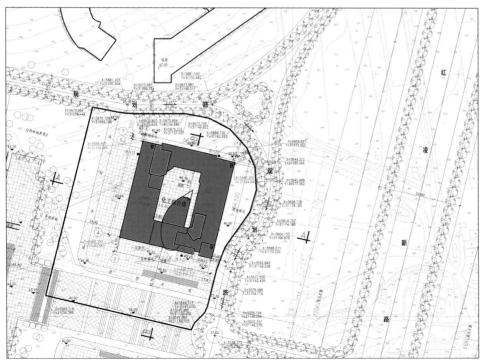

Site Plan 总平面图

Architectural Design

Rational rhythm of unified column width reflects a kind of rhythmic beauty and forms the cultural traits of this campus building. The air chute which connects each floor on the exterior wall looks like a pilastrade, forming an upward tendency. And inserted H-shaped beam steel emphasizes the vertical visual sense.

建筑设计

建筑统一的柱网开间产生理性的韵律，体现了一种节奏美感，形成校园建筑的文化特质。居于建筑外墙、贯通各层的通风道像一排壁柱，形成向上的趋势。中间加入的工字钢构件，强调了纵向的视觉感。

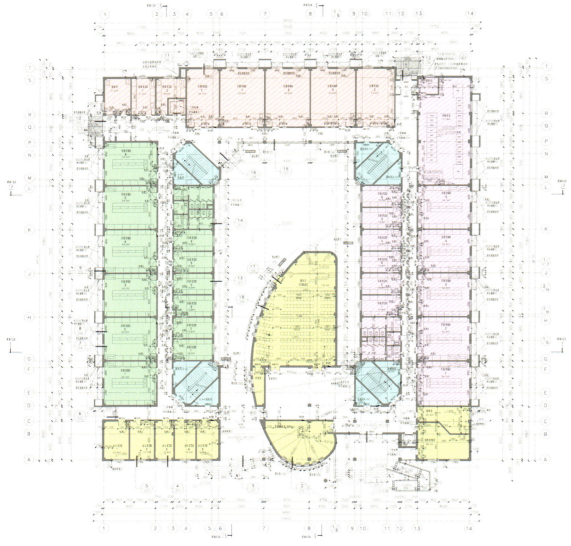

First Floor Plan 首层平面图

Landscape Design

The square in the south has a large area of green space; east and west sides boast terrain-friendly green landscape; looking northward through the gap of the laboratory building, one can see the mountain scenery. In general, this complex building is surrounded by a beautiful and pleasant environment.

景观设计

综合楼南面广场设有大面积绿地；东、西两侧有结合自然地形的绿化景观；向北透过化工实验楼的间隙，可看到校区北面山景；四周环境优美、怡人。

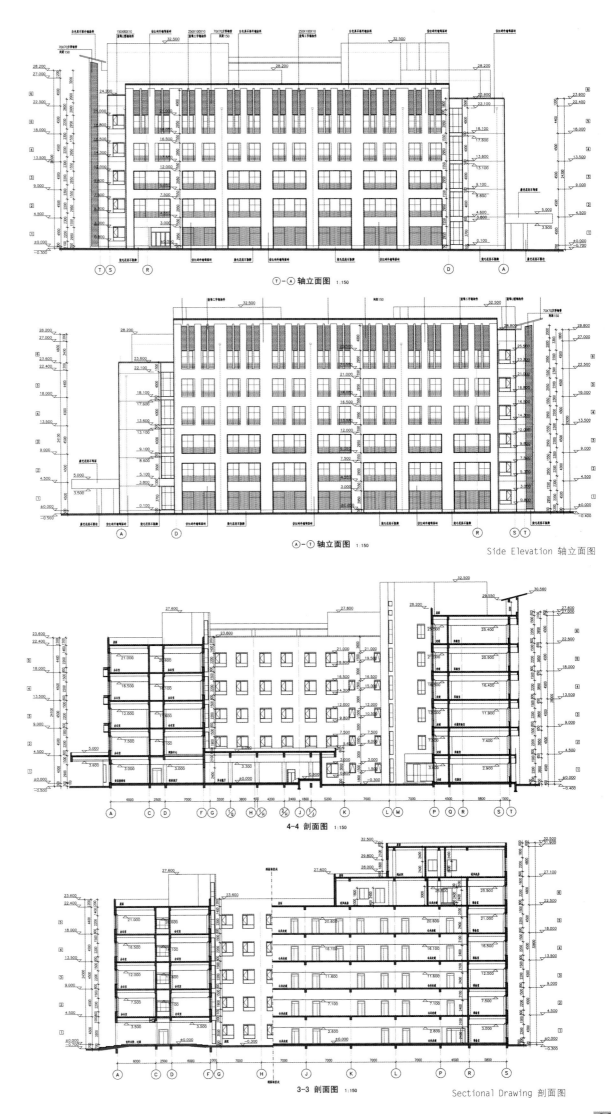

Side Elevation 轴立面图

Sectional Drawing 剖面图

KEY WORDS
关键词

Symbolic Building 形象建筑
Concise Shape 造型简洁
Generous Volume 体量大方

Erdos No.1 Middle School Information Center
鄂尔多斯市第一中学图文信息中心

FEATURES
项目亮点

As a symbolic building in Erdos, this information center introduces contrast of void and solid to building façade, where the glass curtain wall is largely used to interplay with stone wall.

图文信息中心作为鄂尔多斯一中的中心标志性建筑，建筑立面造型采用虚实对比，大面积的玻璃幕墙与石材墙面结合。

Location: Erdos, Inner Mongolia, China
Architectural Design: Tianjin University Research Institute of Architectural Design & Urban Planning
Floor Area: 38,800 m²
Total Floor Area: 37,700 m²
Plot Ratio: 1.03

项目地点：中国内蒙古自治区鄂尔多斯市
设计单位：天津大学建筑设计研究院
建筑面积：38 800 m²
总用地面积：37 700 m²
容积率：1.03

Overview

The building is divided into Area A and Area B, and both of them cover a rectangular plot of land and are connected by an exhibition hall. The main part has one floor underground and five floors overground. The size of the column grid is 9.0 m × 9.0 m.

项目概况

本建筑分为A区和B区，A区和B区均为矩形平面，中间以展厅连接。主体建筑地下一层，地上五层，柱网尺寸主要采用9.0米×9.0米。

经济技术指示
总用地面积：35 9540m²
总建筑面积：155 400m²
其中 教学楼 51 600m²
体育馆 8 200m²
宿舍：48 000m²
食堂：9 100m²
图文中心：38 500m²
建筑密度：11.4%
容积率：0.43
绿化率：39%

Site Plan 总平面图

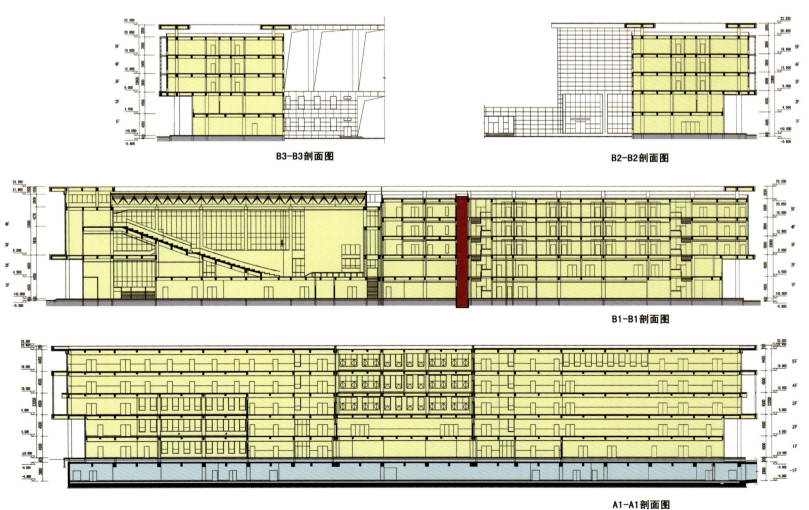

B3-B3剖面图

B2-B2剖面图

B1-B1剖面图

A1-A1剖面图

Sectional Drawing 剖面图

First Floor Plan 首层平面图

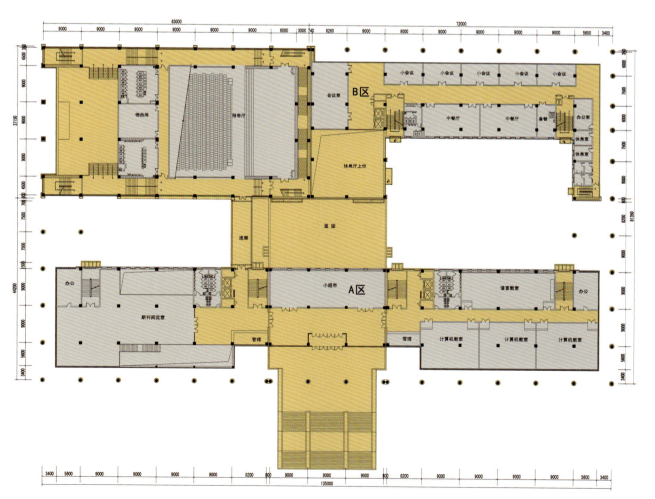

Second Floor Plan 二层平面图

Façade Design

As a symbolic building in Erdos, this information center introduces contrast of void and solid to building façade, where the glass curtain wall is largely used to interplay with stone wall. The glass curtain wall is composed of point-supported glass and Low-E hollow toughened glass, which meet both the lighting need and energy-saving requirement. The main wall is composed of granite material, which shapes a sharp contrast with the glass curtain wall. In terms of the fifth façade, granite stone roof is simple, elegant and contemporary. No wonder that the information center is a symbolic building in Erdos.

立面设计

图文信息中心作为鄂尔多斯一中的中心标志性建筑，建筑立面造型采用虚实对比，大面积的玻璃幕墙与石材墙面结合。玻璃幕墙主要采用点式玻璃，低辐射中空钢化玻璃，既满足采光需要又能达到节能要求。主墙面为花岗岩石材，与玻璃幕墙形成强烈的虚实对比。设计中考虑了第五立面，屋顶采用花岗岩石材屋面，造型简洁、大方、富有现代感，同时也成为鄂尔多斯市的形象建筑。

KEY WORDS 关键词

Elevated Ground Floor 底层架空
Accessible Roof 上人屋面
Widened Corridor 廊道加宽

Hengdan School in Wenxian, Gansu
甘肃文县横丹学校

FEATURES 项目亮点

A part of the building roof is accessible roof which is connected with the ground by ramp, collectively creating a three-dimensional and multi-level outdoor and semi-outdoor activity space system.

本方案将部分建筑屋顶设计为上人屋面，并通过坡道与地面场地联系在一起，共同形成一个立体的、多层次的室外和半室外活动空间体系。

Location: Longnan, Gansu, China
Architectural Design: School of Architecture, Tsinghua University
 An-design Architects
Land Area: 6,700 m²
Total Floor Area: 5,451 m²

项目地点：中国甘肃省陇南市
设计单位：清华大学建筑学院、北京华清安地建筑设计事务所有限公司
占地面积：6 700 m²
总建筑面积：5 451 m²

Overview

New Hengdan School includes five parts: teaching building, dormitory building, canteen, latrine and gate house. A serious shortage of land is the most important challenge the construction faces. In order to offer more outdoor space, the building groups are arranged along the side line of the land to leave plenty of center space as much as possible. Meanwhile, the bottom of the teaching building is elevated so that both sides divided by the building are linked. In addition, the corridor to the inner side of the building is widened properly and forms a semi-outdoor space together with galleries, providing a sheltered stage for the students. Besides, a part of the building roof is accessible roof which is connected with the ground by ramp, collectively creating a three-dimensional and multi-level outdoor and semi-outdoor activity space system.

项目概况

新横丹学校包括教学楼、宿舍楼、食堂、公厕、门房五部分。用地严重不足是学校重建面临的首要问题。为了提供更多的室外场地，建筑群沿用地周边布置，尽量多留出中间场地，同时教学楼底层部分架空，使被分割在建筑两侧的场地彼此贯通。另外，教学楼内侧走道适当加宽，结合廊道和廊架形成半室外空间，在有限的条件下，为学生提供了一个可以遮风避雨的半室外活动场所。再有，本方案将部分建筑屋顶设计为上人屋面，并通过坡道与地面场地联系在一起，共同形成一个立体的、多层次的室外和半室外活动空间体系。

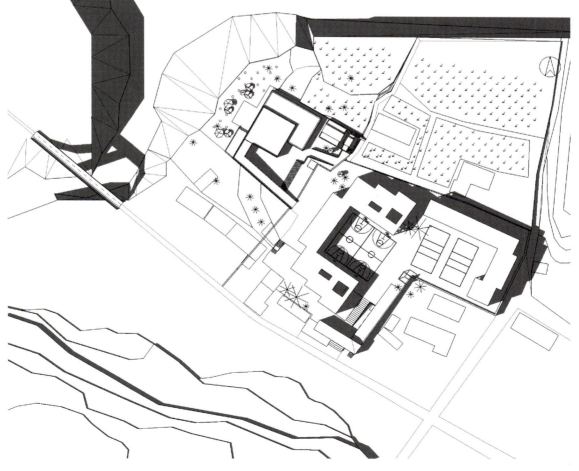

Site Plan 总平面图

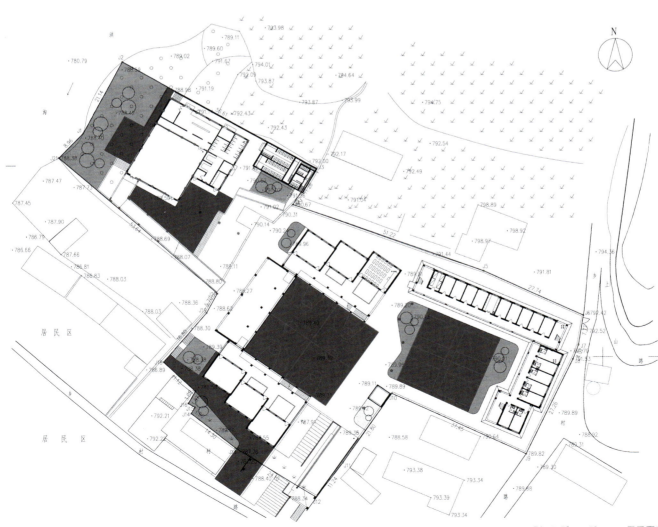

First Floor Plan 一层平面图

East Elevation 东立面图

West Elevation 西立面图

South Elevation 南立面图

Sectional Drawing 剖面图

KEY WORDS	**Elegant Shape** 造型优美
关键词	**Unique Style** 独具特色
	Ecological and Environment-friendly 生态环保

Jiang Nan Shui Du Middle School
江南水都中学

FEATURES
项目亮点

The slope roof of the canteen is designed to be a garden, which serves as the extension of the green playground. It is a natural stand, an outdoor theater and a space for relaxation and gathering.

将食堂设计成绿色的斜屋面花园，作为绿茵操场的延伸，既是自然天成的绿色看台，也是露天剧场、学生休闲聚会的空间，斜屋面上三三两两的学生聚会将成为水都中学的特色之一。

Location: Fuzhou, Fujian, China
Architectural Design: Fujian Provincial Institute of Architectural Design and Research
Land Area: 35,903.2 m^2
Total Site Area: 6,726.3 m^2
Total Floor Area: 19,516.7 m^2
Floor Area aboveground: 18,583.5 m^2
Floor Area underground: 933.2 m^2
Building Density: 18.7%
Greening Rate: 31.3%
Plot Ratio: 0.518

项目地点：中国福建省福州市
设计单位：福建省建筑设计研究院
用地面积：35 903.2 m^2
总占地面积：6 726.3 m^2
总建筑面积：19 516.7 m^2
地上建筑面积：18 583.5 m^2
地下建筑面积：933.2 m^2
建筑密度：18.7%
绿地率：31.3%
容积率：0.518

Overview

The new campus is located on the east of Youxizhou Bridge's southern overpass. To present the passing-by drivers with impressive facades and shield the teaching area from the noises, the architects have built the teaching facilities 80 m east from the road and set the playground near by it.

项目概况

新校区位于尤溪洲大桥南立交的东侧，该地段噪声很大，同时地段的重要性也要求建筑立面上要给穿梭而过的车辆留下深刻的印象。因此设计中让教学区向东退 80 m 以上，把嘈杂留给运动区域。

Architectural Design

The slope roof of the canteen is designed to be a garden, which serves as the extension of the green playground. It is a natural stand, an outdoor theater and a space for relaxation and gathering. The main entrance is set on the east main road and connects with the living area in the southeast, while the sub entrance is built on the west urban road. In the southeast corner of the campus is the small stadium which is connected with the teaching buildings by a corridor. Art gallery is designed along the corridor to enhance the artistic atmosphere of the campus.

建筑设计

将食堂设计成绿色的斜屋面花园，作为绿茵操场的延伸，既是自然天成的绿色看台，也是露天剧场、学生休闲聚会的空间，斜屋面上三三两两的学生聚会将成为水都中学的特色之一。主入口设于东侧小区干道上，安全且便于与东南侧居住区联系；在西侧城市干道上设学校的次入口。小型体育馆布置在地块的东南角，与教学区尽量拉开距离，用风雨廊相连，使整体建筑朝向尤溪洲大桥的面拉长，长廊边设艺术园地，增加校园艺术气息。

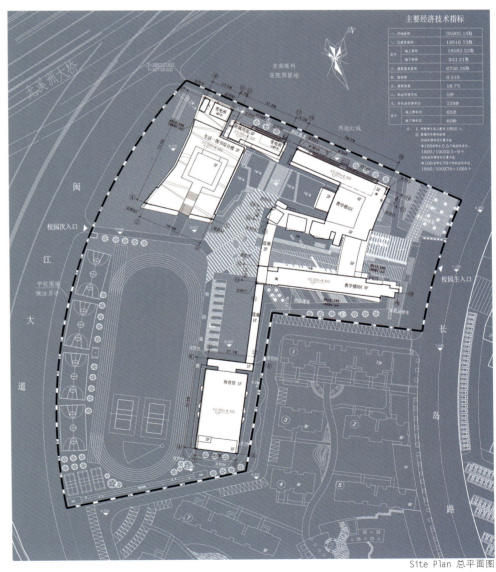

Site Plan 总平面图

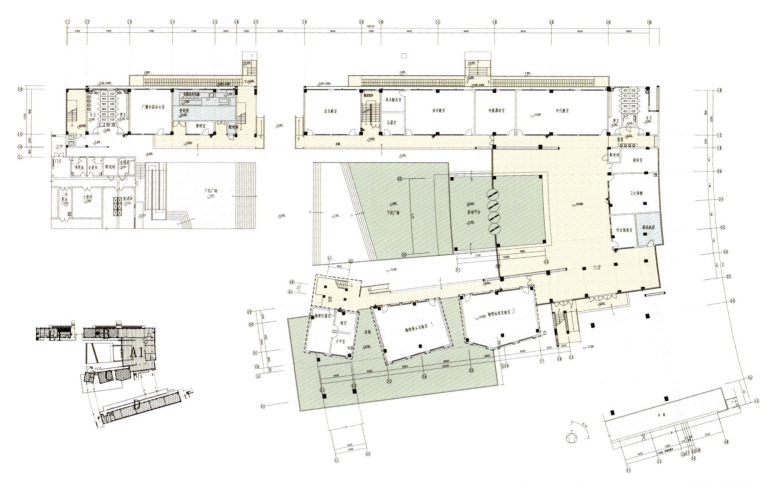

First Floor Plan of Teaching Area 教学区一层平面图

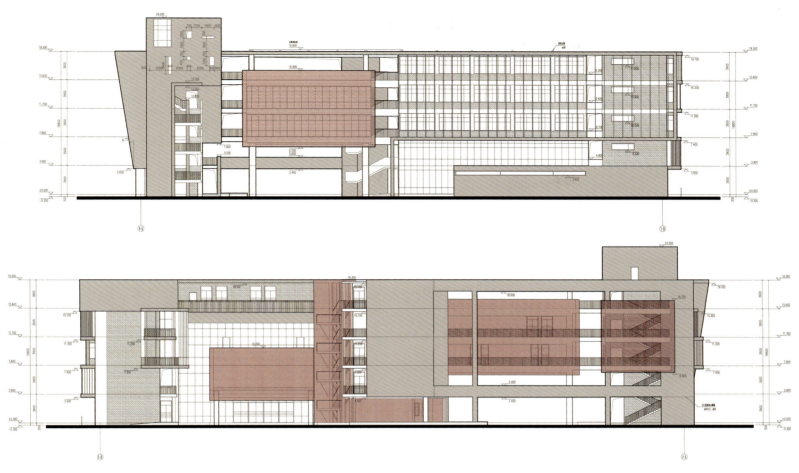

Elevation of Teaching Area 教学区展开立面图

KEY WORDS 关键词	Harmonious Colors 色调和谐
	Elegant Style 简洁大气
	Reasonable Organization 动静分区

New Campus of Jiangyin High School
江苏省江阴高级中学新校区

FEATURES 项目亮点

Different functions are organized reasonably; diversified courtyard spaces meet the requirements for teaching, learning and communication. It is an ideal place for working and studying, and it has become a new landmark of Jiangyin City.

规划合理动静分区设计，多样的院落空间给师生带来舒适的交流学习空间，充分满足了教学需要，给校院带来了良好的学习、工作环境，同时也成为江阴城市的一个新标志。

Location: Jiangyin, Jiangsu, China
Architectural Design: Jiangyin Architectural Design & Research Institute Co.,Ltd.
Planned Total Floor Area: 11,000 m²
Plot Ratio: 0.74

项目地点：中国江苏省江阴市
设计单位：江阴市建筑设计研究院有限公司
规划总建筑面积：11 000 m²
容积率：0.74

Overview

As a modern education center to show the culture of a school and the vitality of a city, the new campus fully embodies the cultural atmosphere and education idea of Jiangyin High School. Different functions are organized reasonably; diversified courtyard spaces meet the requirements for teaching, learning and communication. It is an ideal place for working and studying, and it has become a new landmark of Jiangyin City.

项目概况

江苏省江阴高级中学新校区作为展现学校文化风采和当代城市活力的现代教育中心，该设计充分展现了江阴高级中学的人文气息和教育理念。规划合理，动静分区设计，多样的院落空间给师生带来舒适的交流学习空间，充分满足了教学需要，给校院带来了良好的学习、工作环境，同时也成为江阴城市的一个新标志。

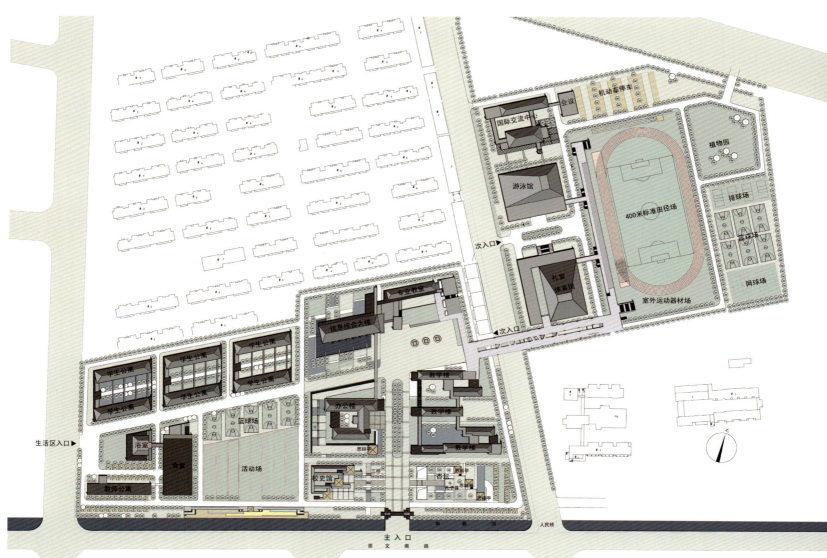

Site Plan 总平面图

KEY WORDS 关键词

Open Ground Floor 底层架空
Cultural Atmosphere 人文精神
Elegant Appearance 外形优美

The North No.2 Liberal Arts Building, Nanjing University of Aeronautics and Astronautics (NUAA) Jiangjun Road East Campus

南航将军路校区东区北 2 号文科楼

FEATURES 项目亮点

The ground floor is open to form an internal courtyard which provides a public space for teachers and students, and dialogues with the campus environment.

建筑通过底层架空的"大空间"形成围合内部"院"空间为院系间提供了有趣味的共享空间,与整个校园环境形成对话。

Location: Nanjing, Jiangsu, China
Architectural Design: Nanjing Yangtze River Urban Architectural Design Co.,Ltd.
Land Area: 3,600 m²
Total Floor Area: 13,600 m²
Building Height: 19.95 m
Plot Ratio: 0.26

项目地点:中国江苏省南京市
设计单位:南京长江都市建筑设计股份有限公司
建筑用地面积:3 600 m²
总建筑面积:13 600 m²
建筑高度:19.95 m
容积率:0.26

Overview

The Jiangjun Road East Campus of Nanjing University of Aeronautics and Astronautics (NUAA) is located in Jiangning Economic and Technological Development Zone, 13 km away from its Ming Imperial Palace campus. Reaching Shengtaixi Road on the north, Changtai Road on the south, Jinghuai Road on the east, and the airport expressway on the west, the east campus is just across a road from the already built west campus. The No.2 Liberal Arts Building in the east campus is the teaching building for the College of Foreign Languages and College of Humanities.

项目总概况

南航将军路校区东区校园位于南京市江宁经济技术开发区,距明故宫校区约 13 km。北起胜太西路,南至长泰路,东接静淮路,西抵机场高速公路,与已建成的西区校园仅一路之隔。其东区北 2 号文科楼实为外语学院和人文学院的教学楼。

Architectural Design

As the teaching building for these two colleges, it features the atmosphere of humanities, arts research and education. With simple and natural style, the building keeps harmonious with the other buildings and facilities. The ground floor is open to form an internal courtyard which provides a public space for teachers and students, and dialogues with the campus environment.

建筑设计

作为外语学院、人文学院的文科楼，该建筑将人文精神、艺术研究和教育结合在一起，通过质朴、自然的风格，保持建筑与校园内其他建筑的协调与统一。建筑通过底层架空的"大空间"形成围合内部"院"空间为院系间提供了有趣味的共享空间，与整个校园环境形成对话。

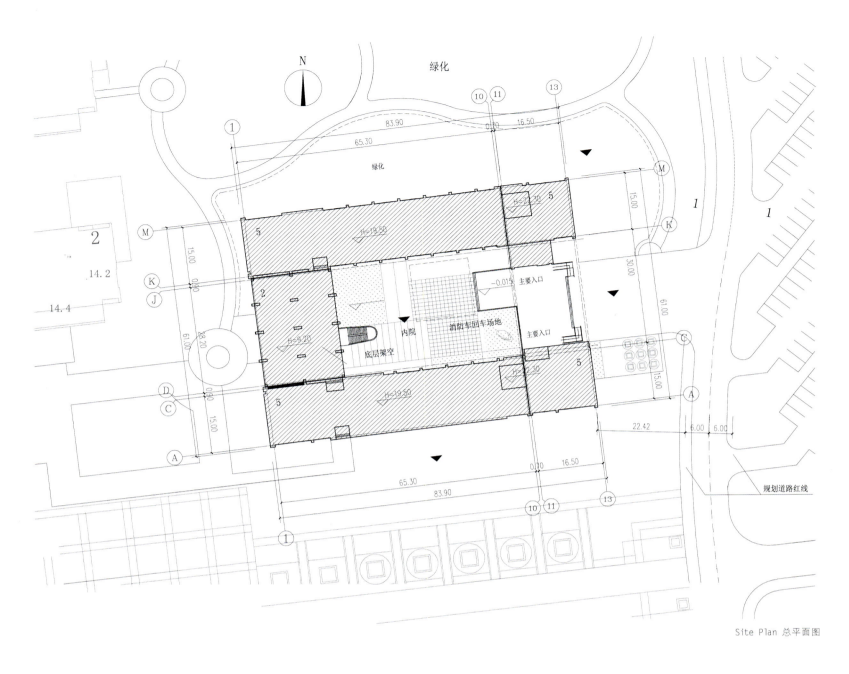

Site Plan 总平面图

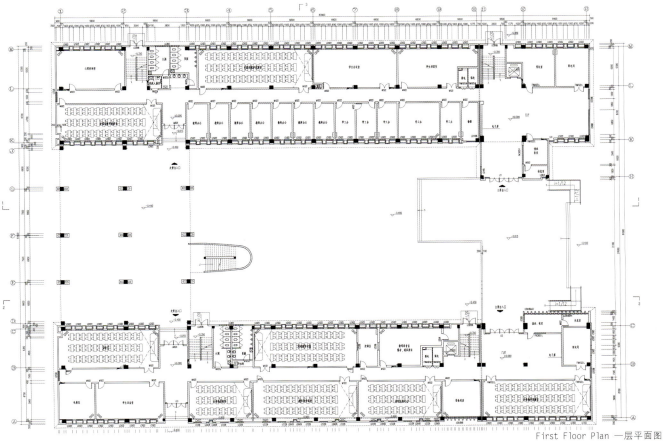

First Floor Plan 一层平面图

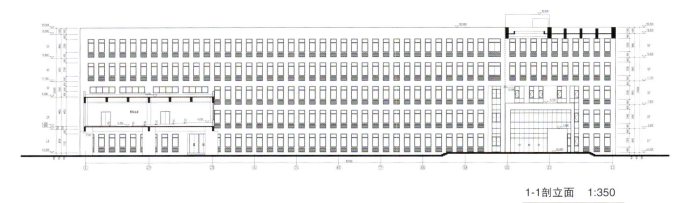

1-1剖立面　1:350

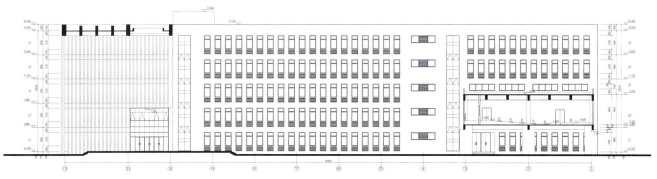

Sectional Drawing 剖面图

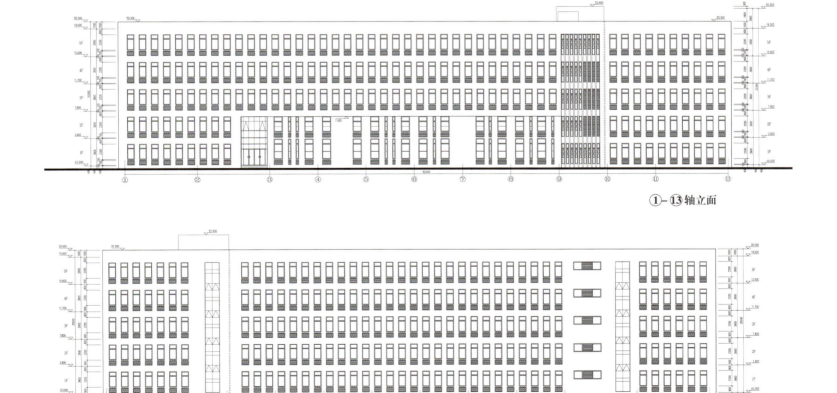

①-⑬轴立面

⑬-①轴立面

Side Elevation 轴立面图

KEY WORDS 关键词

Imposing 大气恢宏

Unique Appearance 外观独特

Bright Color 色调鲜明

Century Hall of Tsinghua University
清华百年会堂

FEATURES 项目亮点

As a new landmark for Tsinghua University, the century hall is a complex for conference, performance and exhibition.

清华大学百年会堂是集会议、表演、展览于一体的综合性建筑,是清华大学新百年的标志性建筑。

Location: Haidian, Beijing, China
Architectural Design: Architectural Design and Research Institute of Tsinghua University
Total Floor Area: 42,950.5 m²
Floor Area (aboveground): 31,720 m²
Floor Area (underground): 11,230.5 m²
Plot Ratio: 1.26
Greening Rate: 45%

项目地点:中国北京市海淀区
设计单位:清华大学建筑设计研究院
总建筑面积:42 950.5 m²
地上建筑面积:31 720 m²
地下建筑面积:11 230.5 m²
容积率:1.26
绿化率:45%

Overview

Located in the campus of Tsinghua University, Haidian District of Beijing, the century hall is situated at the northeast corner where two main roads cross. The site reaches the south-north main road on the west, the east-west main road on the south, the Main Building West Road on the east, and the No.3 Teaching Area on the north.

项目概况

清华大学百年会堂位于北京市海淀区清华大学校园内,清华大学东西主干道与南北主干道交叉的东北角。地段西至校南北干道,南至校东西干道,东至主楼西路,北至第三教学楼区南路。

Planning

As a new landmark for Tsinghua University, the century hall is a complex for conference, performance and exhibition. It is composed of three parts: new Tsinghua Auditorium, Mong Man-wai Concert Hall, Tsinghua University History Museum and other supporting facilities. The new Tsinghua Auditorium is a supersized class-B theater which has 2,011 seats for large-scale performances and conferences, and the 500-seat Mong Man-wai Concert Hall is a small class-B theater for small shows and concerts.

规划布局

清华大学百年会堂是集会议、表演、展览于一体的综合性建筑,是清华大学新百年的标志性建筑。清华大学百年会堂包括三个主要部分:新清华学堂、蒙民伟音乐厅、清华大学校史馆以及附属设施。新清华学堂为超大型乙等剧场,共2 011座,能够满足大型演出、会议的需要;蒙民伟音乐厅为乙等小型剧场,共500席,能满足实验剧和小型音乐会演出需要。

KEY WORDS 关键词

Integration and Inheritance 整合延续

Combination of New and Old 新旧融合

Cultural Connotation 文化内涵

Academy for International Business Officials (AIBO), MOFCOM
商务部国际商务官员研修学院

FEATURES 项目亮点

The architects made some changes to the original architectural structures and added some new functions. New construction is well combined with the existing ones to show the history and culture of the building complex.

对原有建筑功能布局进行巧妙的改造与利用，恰如其分地将新功能穿插安排其间，展示新建部分特色的同时注意传递原有建筑的历史文化信息。

Location: Changping, Beijing, China
Architectural Design: Beijing Institute of Architectural Design
Land Area: 398,229.12 m²
Total Floor Area: 27,173 m²
Plot Ratio: 0.4
Completion: 2009

项目地点：中国北京市昌平区
设计单位：北京市建筑设计研究院有限公司
用地面积：398 229.12 m²
总建筑面积：27 173 m²
容积率：0.4
竣工时间：2009年

Overview

Located in Beiqijia Town of Changping District, Beijing, inside the training center of the Minister of Commerce, PRC., the project is renovated and extended based on the old Asia-Pacific Center.

项目概况

该项目位于北京市昌平区北七家镇商务部培训中心院内，是在原亚太中心项目基础上改扩建而成。

Design Idea

The architects took advantage of the original architectural structures and made some changes to add some new functions. New construction is well combined with the existing ones to show the history and culture of the building complex.

设计理念

对原有建筑功能布局进行巧妙的改造与利用，恰如其分地将新功能穿插安排其间，注重新老部分的和谐统一，在展示新建部分特色的同时注意传递原有建筑的历史文化信息。

Site Plan 总平面图

First Floor Plan of Health Center 康体中心首层平面图

First Floor Plan of Students Apartment 学员宿舍首层平面图

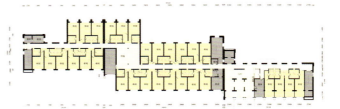

First Floor Plan of Expert Building 专家楼首层平面图

First Floor Plan of Cultural & PE Building 文体楼首层平面图

South Elevation of Cultural & PE Building 文体楼南立面图

South Elevation of No1 Apartment 1号宿舍楼

Second Floor Plan of Students Dining Hall 学员餐厅二层平面图

First Floor Plan of Students Dining Hall 学员餐厅首层平面图

Second Floor Plan 二层平面图

First Floor Plan 首层平面图

West Elevation of Students Dining Hall 学员餐厅西立面图

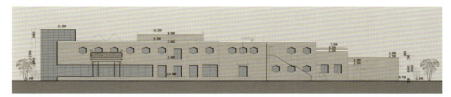
South Elevation of Students Dining Hall 学员餐厅南立面图

North Elevation of Students Apartment 学员宿舍北立面图　　East Elevation of Students Apartment 学员宿舍东立面图

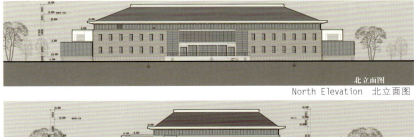
North Elevation 北立面图

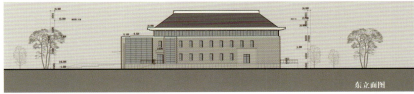
East Elevation 东立面图

South Elevation 南立面图

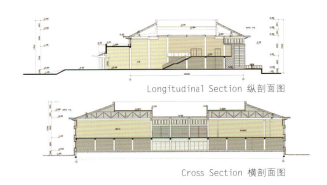
Longitudinal Section 纵剖面图

Cross Section 横剖面图

067

Architectural Design

The newly built construction has complemented the existing buildings. Symmetrical library block is preserved and the sequence along the central axis is further enhanced. Buildings in the west and east stand in balance and are connected together with the main building by the old expert apartments and newly built dormitory.

建筑设计

新建部分充分协调原有布局，使之完整均衡，疏密有致。保持原图书馆方正对称的形态，并进一步加强了中轴线的序列层次。东西侧建筑群体量均衡，原专家公寓与新建学院宿舍采用线性内院式围合布局，很好地将东西两侧建筑与中间主体相连，使建筑群体更趋完整协调。

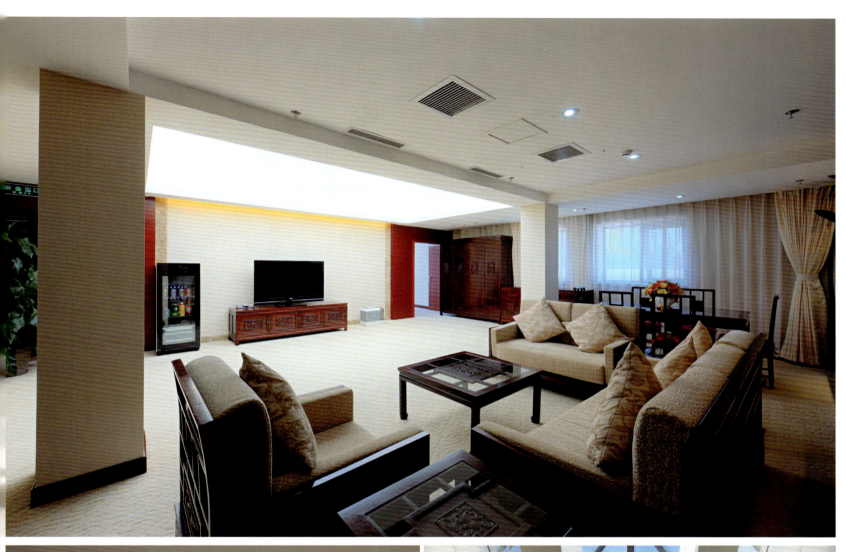

KEY WORDS 关键词

Concrete Facade 混凝土立面
Courtyard Space 庭院空间
Harmonious Environment 环境融合

New Campus of Shenzhen Institute of Information and Technology
深圳信息职业技术学院迁址新建工程

FEATURES 项目亮点

To create a landscape environment, the south campus is connected with the north one by water feature in the center, while the east campus is built with high density dormitory buildings.

设计旨在体现出山水校园的整体环境，依据基地情况，南北校区以中央水体为中心，与山水环境融为一体；东区，则在山水校园中体现高密度的城市化设计。

Location: Shenzhen, Guangdong, China
Architectural Design: The Institute of Architecture Design & Research of Shenzhen University
Total Floor Area: 267,735 m²
Completion: 2011

项目地点：中国广东省深圳市
设计单位：深圳大学建筑设计研究院
总建筑面积：267 735 m²
竣工时间：2011 年

Overview

The new campus of Shenzhen Institute of Information and Technology (used as the Athlete's Village for the 26th Summer Universiade, Shenzhen), boasts a total floor area of 267,735 m². The programs include student dormitories, canteen, staff quarters, school clinic and other living facilities. It has also preserved the Games-time facilities such as the service center, logistics rooms, commercial center and some characteristic Universiade buildings. In addition, there are also some auxiliary rooms, underground parking and open ground floor spaces.

项目概况

深圳信息职业技术学院迁址新建工程（深圳第26届世界大运会运动员村）总建筑面积为267 735 m²，项目主要包括学生宿舍，教工、学生食堂，教工宿舍及其他生活用房，赛时综合服务楼及后勤附属用房和特色建筑，赛时商业中心及大运会特色建筑，校医务室，其他辅助用房，地下停车场，以及架空层。

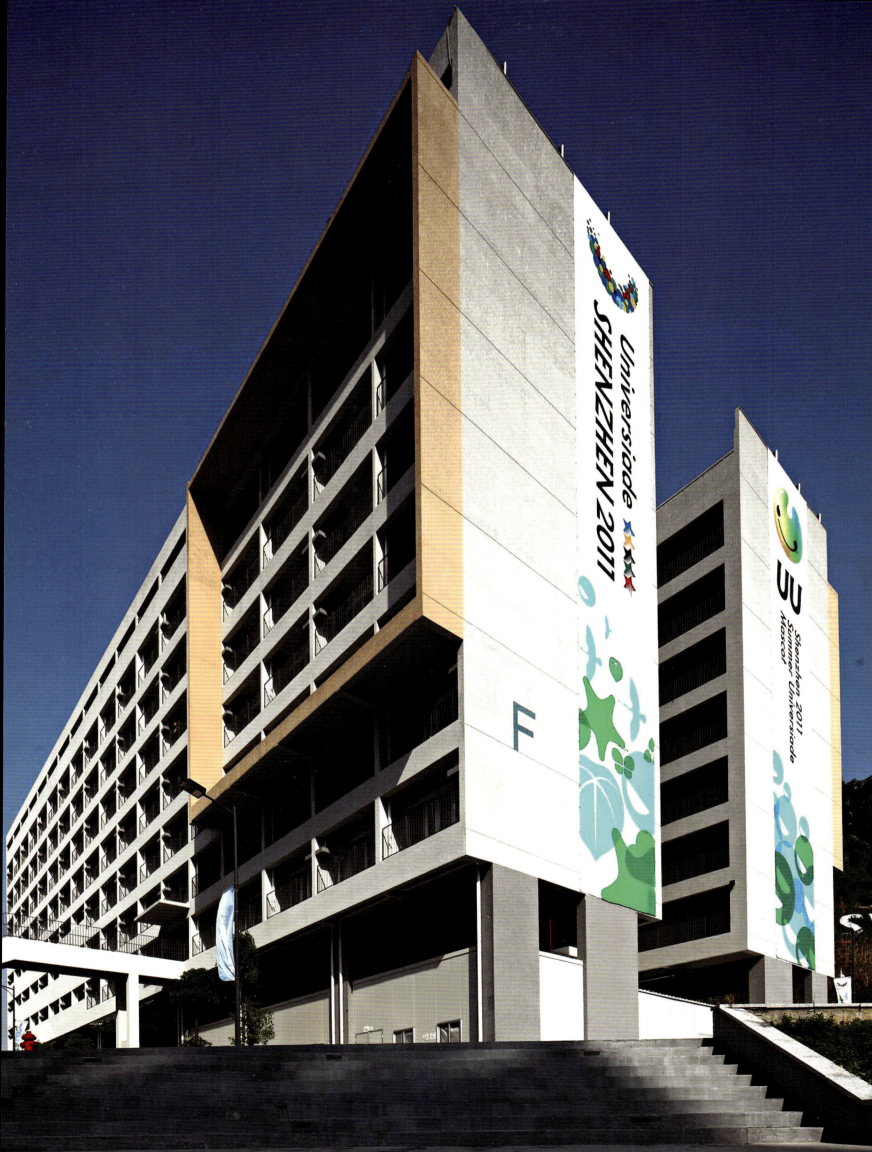

Site Plan 总平面图

Architectural Design

The new campus is designed with the idea to build a landscape environment. The south campus is connected with the north one by the water feature in the center. Relatively lower building density ensures a high quality living and learning environment. Due to limited land area, it implemented high density design and built five 17-storey dormitory buildings in the east. Five high rises are arranged in fold line with south-north orientation and maximum building distance, to get more sunlight and natural wind. Spaces enclosed by buildings are designed as courtyards to optimize the living environment.

建筑设计

新建工程的设计理念希望能够体现出山水校园的整体环境，南北校区以中央水体为中心，整体建设密度相对比较低，较好地与山水环境融合为一体。生活和居住区域，由于可建设用地有限，要在山水校园当中体现高密度的城市化设计。5栋折线状17层高的宿舍区塑造了东校区的总体形象。在用地小、建筑密度大的现实条件下，在尽可能地获得南北朝向的前提下，尽可能拉大楼体间距，以获得更好的采光和通风效果。并通过楼体间的围合形成多个均质的庭院，活跃了宿舍建筑的底层空间，优化学生的生活和居住环境。

KEY WORDS 关键词

Natural and Harmonious 自然和谐
Solemn and Serene 庄严肃穆
Stone Material 石材

Weifang Municipal Party Committee Party School Teaching & Administration Department

潍坊市市委党校教学行政组团

FEATURES 项目亮点

The designers establish the concept of "hidden rather than flaunt, open rather than closed, solemn rather than informal", trying to provide the building with a kind of humility and deliberate attitude to get along with the natural environment.

设计确立了"宜隐不宜显、宜敞不宜闭、宜庄不宜谐"的理念，意图使建筑以一种谦逊从容的姿态与自然环境相容。

Location: Weifang, Shandong, China
Architectural Design: The Architectural Design & Research Institute of Zhejiang University
Planning Land Area (Main Campus): 476,002.38 m²
Total Floor Area (Teaching & Administration Department): 22,883.6 m²
Building Density (Main Campus): 8.42%
Plot Ratio (Main Campus): 0.212
Greening Ratio (Main Campus): 70.74%
Completion: 2010

项目地点：中国山东省潍坊市
设计单位：浙江大学建筑设计研究院
规划用地面积（总校区）：476 002.38 m²
总建筑面积（教学行政组团）：22 883.6 m²
建筑密度（总校区）：8.42%
容 积 率（总校区）：0.212
绿 地 率（总校区）：70.74%
竣工时间：2010 年

Overview

The project base is surrounded by mountain and water. Original ecological slope, gully, wetland, water, pine forest inside the field constitute superior natural environment and resources. "How to implant the building into the base?" and "what kind of temperament should the building have?" are the designers' questions to think about at the beginning of the design. Based on the scene investigation and rational analysis, the designers establish the concept of "hidden rather than flaunt, open rather than closed, solemn rather than informal", trying to provide the building with a kind of humility and deliberate attitude to get along with the natural environment.

项目概况

项目基地临山面水。场地内原生态的坡地、冲沟、湿地、水面、松树林，构成了优越的自然环境资源。"如何使建筑植入基地？""建筑应该具有怎样的气质？"是设计之初的自我提问。结合现场勘察与理性分析，设计确立了"宜隐不宜显、宜敞不宜闭、宜庄不宜谐"的理念，意图使建筑以一种谦逊从容的姿态与自然环境相容。

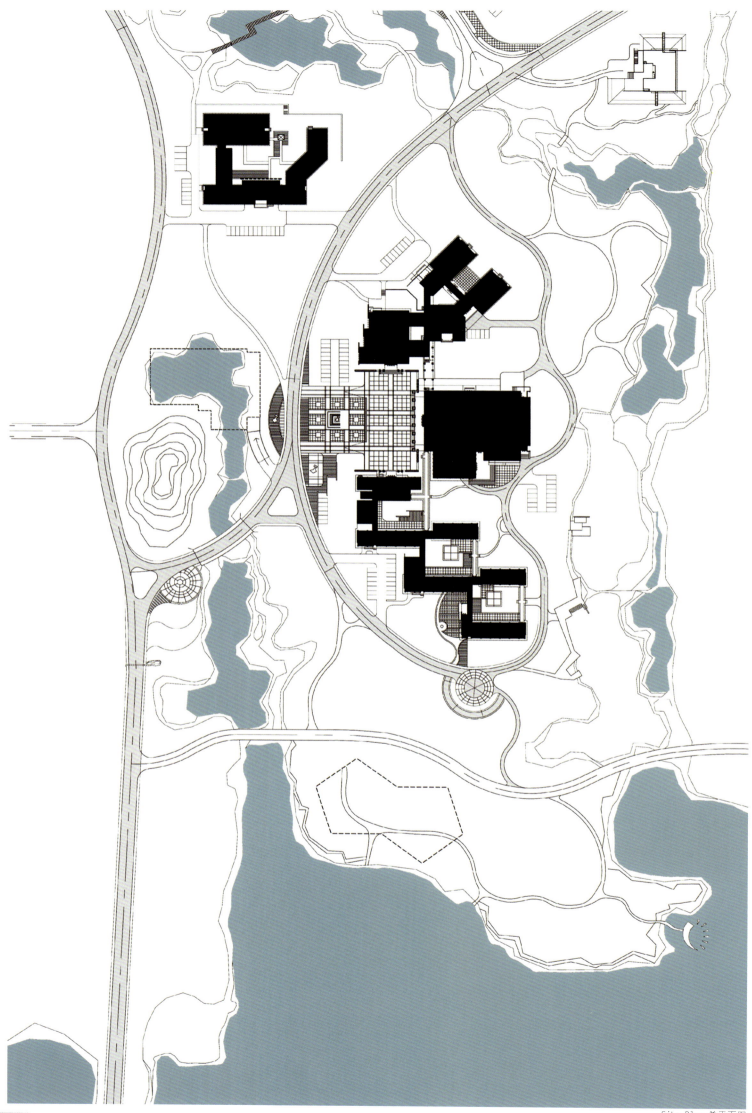

Site Plan 总平面图

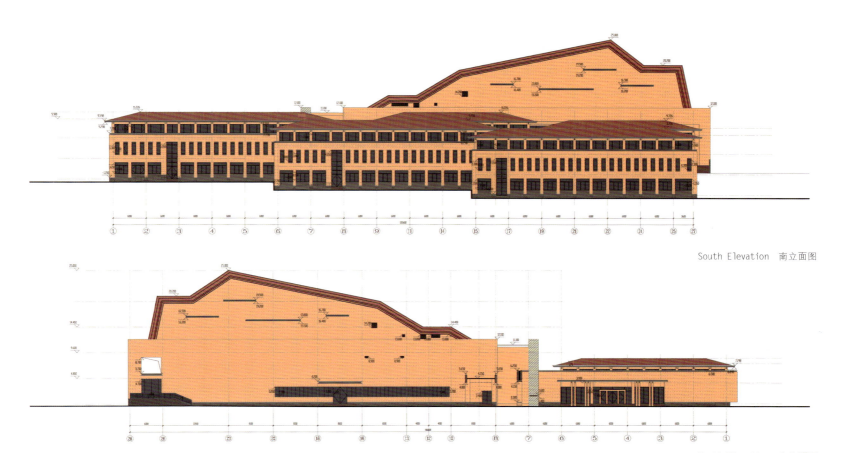

South Elevation 南立面图

North Elevation 北立面图

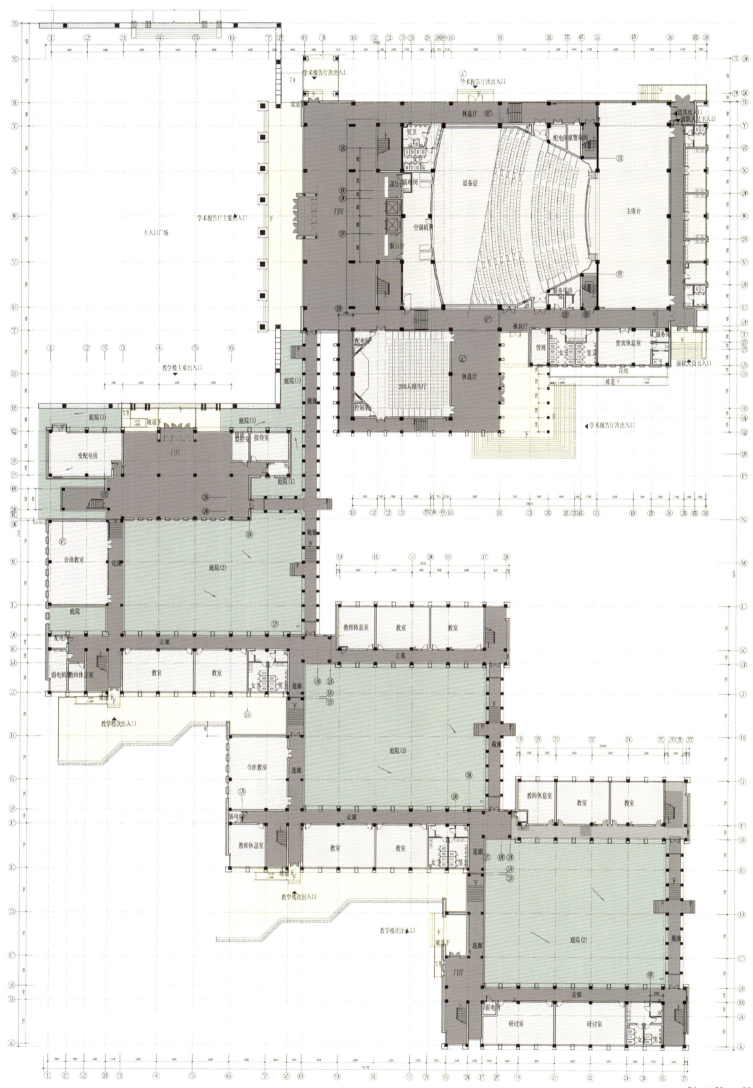

First Floor Plan 一层平面图

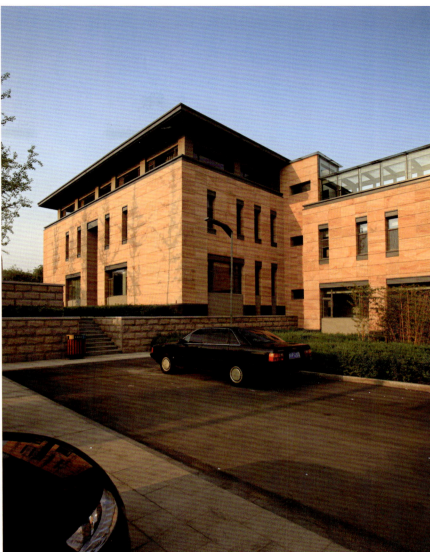
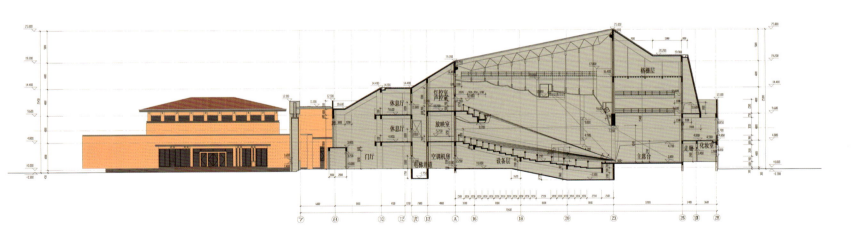
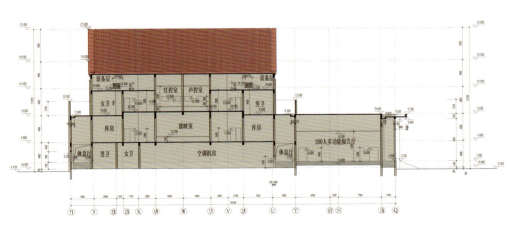

Sectional Drawing 剖面图

KEY WORDS 关键词	**Jiangnan Style** 江南风格
	Cultural Characteristic 人文特色
	Space 场所空间

Wuxi City College of Vocational Technology New Campus

无锡城市职业技术学院新校区

FEATURES 项目亮点

The integration of the campus architecture and garden style and the combination of natural scenery and geometric landscape not only contain the romantic charm of Jiangnan traditional garden, but also bear the preciseness of modern college building.

将校园建筑与园林风格相融合，自然景观与几何景观结合，既含江南传统园林之神韵，又得现代院校之严谨。

Location: Wuxi, Jiangsu, China
Architectural Design: The Architectural Design & Research Institute of Zhejiang University
Completion: 2010

项目地点：中国江苏省无锡市
设计单位：浙江大学建筑设计研究院
竣工时间：2010年

Overview

Wuxi City College of Vocational Technology New Campus (Phase I) has a total construction area about 140,000 m², including the floor 139,920 m² above the ground and 1,275 m² under the ground. This project consists of the library, teaching building, practical training building, administrative office building, gymnasium, dining room, bathroom, student apartments, employee apartments and training building, etc. The library has 10 layers above the ground and 1 layer underground, which belongs to the secondary high-rise. Others are mainly multi-storey building or low-rise building. The overall project has a land area of 343,221 m², total plot ratio of 0.64, total building density of 15% and a total greening ratio of 39%.

项目概况

无锡城市职业技术学院新校区（一期）总建筑面积约140 000 m²，其中地上建筑面积139 920 m²，地下建筑面积1 275 m²。本案建筑由图书馆、教学楼、实训楼、行政办公楼、体育馆、食堂、浴室、学生公寓、职工公寓、干训楼等主要功能建筑组成。其中图书馆地上10层，地下一层，属二类高层；其他建筑均为多层建筑或低层建筑。可建设用地面积343 221 m²，总容积率0.64，总建筑密度15%，总绿地率39%。

Architectural Design

The integration of the campus architecture and garden style and the combination of natural scenery and geometric landscape not only contain the romantic charm of Jiangnan traditional garden, but also bear the preciseness of modern college building. It seeks rationality in freedom, and realizes order from the nature. Creating a gentle, harmonious and beautiful campus environment and presenting Jiangnan culture characteristics is the objective of this design theme. The building adopts Jiangnan garden color system of black, white and grey to set off itself in the green lake, forming elegant and classical architectural style. The project also strives to create a free, innovative, peaceful and leisurely campus space, so as to achieve an ideal campus atmosphere for study and research.

建筑设计

将校园建筑与园林风格相融合，自然景观和几何景观结合，既含江南传统园林之神韵，又得现代院校之严谨。在自由之中见理性，于自然之中见规矩方圆。创造亲人、和谐、优美的校园环境，体现江南人文特色，是本项目的设计主题。建筑色调采用江南园林建筑的黑白灰色系，掩映于湖水绿化之中，形成优雅、经典的建筑风格。力求塑造出自由、创新、淡定、从容的校园场所空间，获得宁静以致远，心静以治学的整体校园氛围。

Site Plan 总平面图

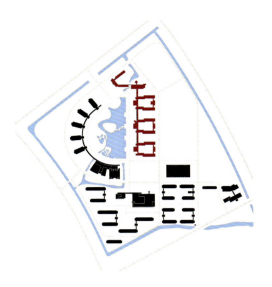

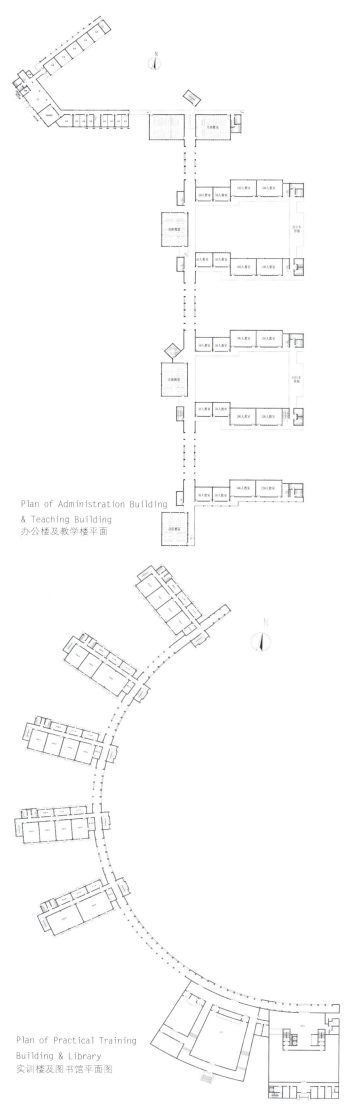

Plan of Administration Building
& Teaching Building
办公楼及教学楼平面

Plan of Practical Training
Building & Library
实训楼及图书馆平面图

湖面看教学楼

KEY WORDS 关键词

Functional Layout 功能布局

Landscape Elements 山水元素

Green Ecology 绿色生态

Central South University Comprehensive Teaching Building
中南大学综合教学楼

FEATURES 项目亮点

The overall design pays attention to the carry and continuity of natural elements, greening and landscape, and harmoniously infuses them into the building.

整体设计注重自然元素的承载和延续性，强调绿化与山水，并将这些和谐地融入建筑之中。

Location: Changsha, Hunan, China
Architectural Design: Hunan Architectural Design Institute
Completion: 2009

项目地点：中国湖南省长沙市
设计单位：湖南省建筑设计院
竣工时间：2009 年

Overview

Central South University Comprehensive Teaching Building is located in the core area of the new campus center which is under planning and construction; it has Changsha Mount Yuelu Scenic Area as the background, links with Jinjiang Road and Houhu Park in the north, faces the Xiangjiang River ecological scenic belt in the east and the second city ring in the west. There is a total construction area of 80,000 m², spreading Block A, B, C, E from north to south; Block A and B in the northern teaching building have a total construction area of 41,717 m², and Block C and E in the southern teaching building area have a construction area of 25,405 m², plot ratio of 0.63, and greening ratio of 52%.

项目概况

中南大学综合教学楼位于规划建设的新校园中心的核心区，以长沙岳麓山风景名胜区为背景，北接城市道路靳江路及后湖公园，东临湘江生态风光带，西临城市二环线。总建筑面积为 80 000 m²，从北到南布置 A、B、C、E 四区，教学楼北楼 A、B 区建筑面积为 41 717 m²，南楼 C、E 区建筑面积为 25 405 m²，容积率为 0.63，绿化率为 52%。

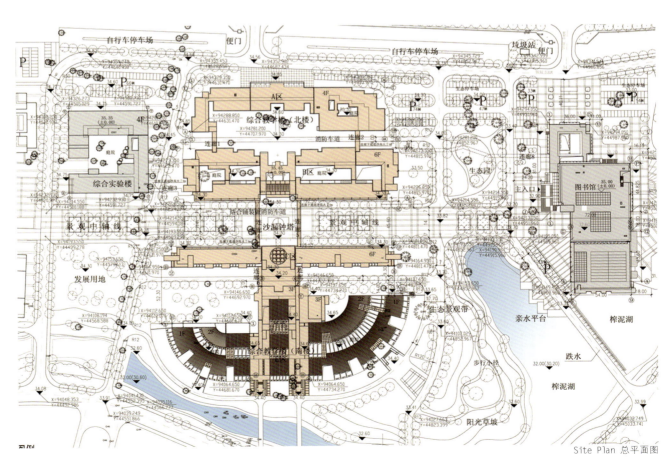

Site Plan 总平面图

Architectural Design

Xiangjiang River and Mount Yuelu infuse ecological vitality into the campus image construction and development. The core area of the new campus organizes each functional block with "one center and two axes", and comprehensive teaching building lies at the intersection of the east-west traffic axis and the north-south landscape axis. The new campus is surrounded by mountains and rivers on three sides, forming the pattern of integrated landscape and buildings. By virtue of superior site conditions, the overall design pays attention to the carry and continuity of natural elements, greening and landscape, and also maintains the relative climate regulation ability while making it become an inseparable part of the building.

建筑设计

湘江与岳麓山为新校区校园形象的塑造和发展注入了生态活力。新校区核心区以"一心二轴"组织各项功能，综合教学楼处于东西交通轴与南北景观轴的交会处，三面山水环绕，形成山水建筑融为一体的格局。凭借优越的场地条件，整体设计上注重自然因素的承载和延续性，强调绿化与山水，在保持其相对的气候调节能力的同时，使其成为建筑不可分割的一部分。

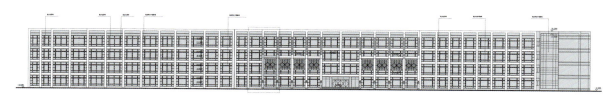

A区 ⑭—① 立面

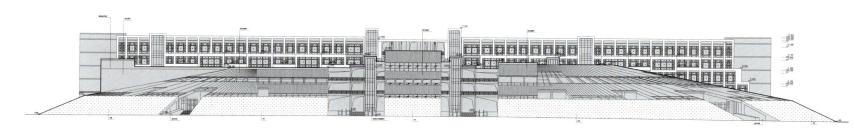

E区 ①—⑬ 立面

Elevation 立面图

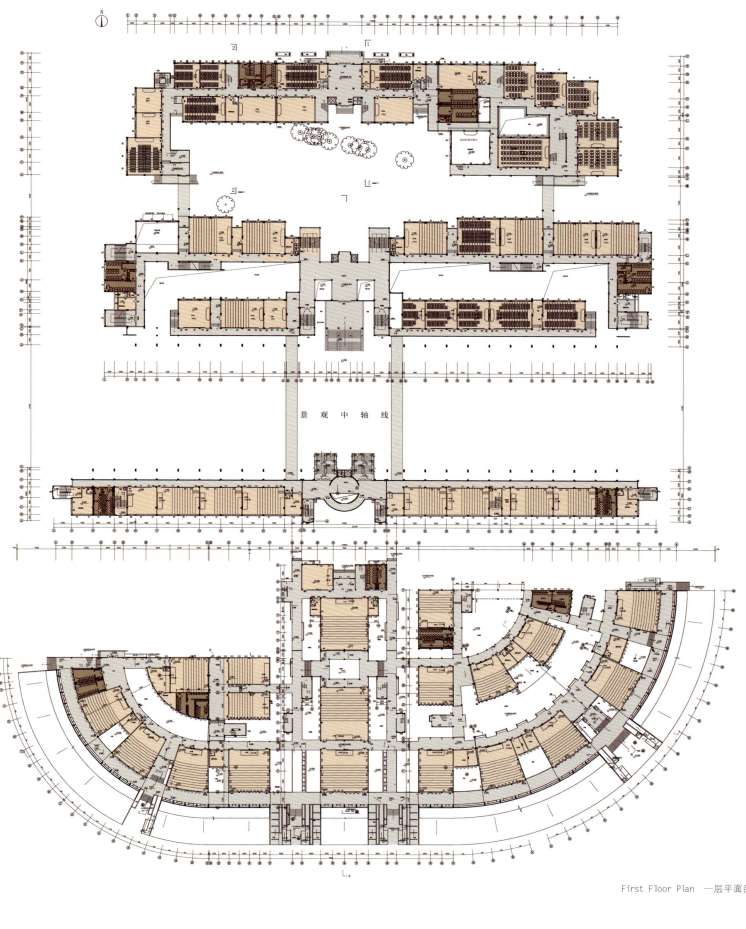

First Floor Plan 一层平面图

B区 3—3剖面　　　　　　　　　　　　　　　　　　　Sectional Drawing 剖面图

E区 Ⓐ–Ⓗ🅐 立面　　　　　　　　　　　　　　　　　Elevation 立面图

KEY WORDS 关键词

Innovative Form 形态新颖
Cultural Connotation 文化内涵
Comprehensive Function 功能综合

International Students Apartment of Peking University
北京大学留学生公寓

FEATURES 项目亮点

Clear conception, innovative form and concise structure highlight the features and functional requirements of an apartment building. Meanwhile, morphological language of the building and the environment are mobilized and shape a vibrant building group.

建筑构思明确，形态新颖而富有哲理，形体简洁而不单调，突出公寓建筑的特征及使用的功能性要求，塑造出了富有生机的建筑群。

Location: Haidian, Beijing, China
Architectural Design: China Architecture Design & Research Group
Floor Area: 131,034 m²
Completion: 2010

项目地点：中国北京市海淀区
设计单位：中国建筑设计研究院
建筑面积：131 034 m²
竣工时间：2010 年

Overview

Located in Zhongguancun, Haidian, Beijing, this project consists of the student apartment, serviced apartment and rooms for office management, and is a comprehensive apartment which integrates accommodation, catering, conference, cultural exchange, office, fitness and recreation, etc.

项目概况

北京大学留学生公寓位于北京市海淀区中关村。项目包括学生公寓、酒店式公寓及服务办公管理用房等，是一个集住宿、餐饮、会议、文化交流、办公休闲、健身娱乐于一体的综合性公寓园区。

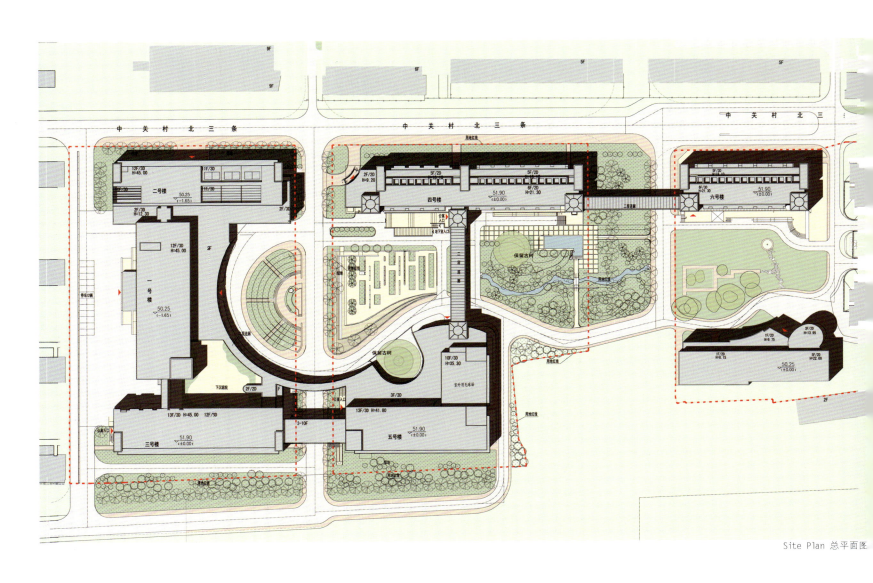

Site Plan 总平面图

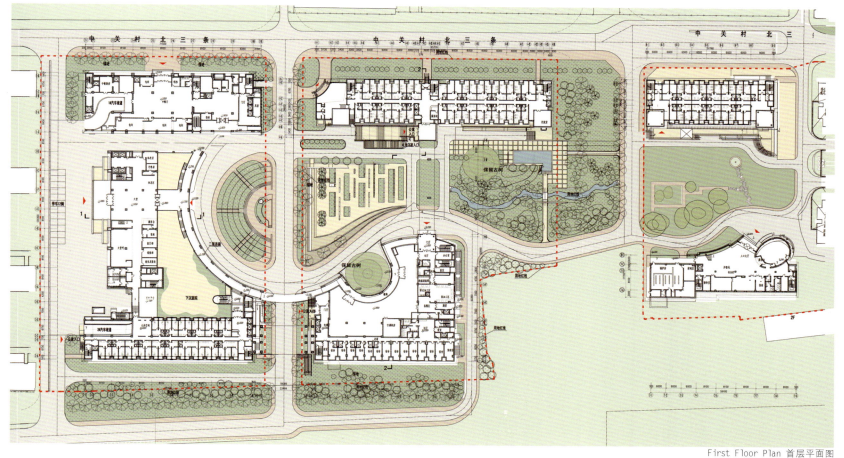

First Floor Plan 首层平面图

Architectural Design

In consideration of terrain features and natural landscape, designers created a building group surrounding a central courtyard. Clear conception, innovative form and concise structure highlight the features and functional requirements of an apartment building. Meanwhile, morphological language of the building and the environment are mobilized and shape a vibrant building group. All the possible elements are used to fully satisfy both the function and the economic requirement. Boasting cultural connotation, humanistic spirit of Peking University and decades of management experience of international students, it is an iconic building group for modern Peking University.

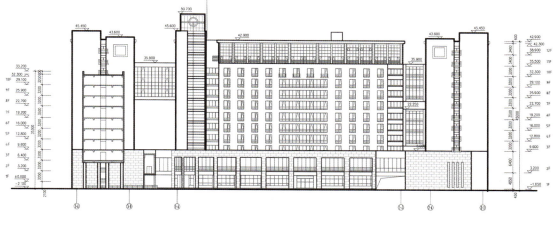

East Elevation 东立面图

建筑设计

北京大学留学生公寓利用地形特征、自然景观，从功能出发形成逐渐升起，合院环抱的建筑群。建筑构思明确，形态新颖而富有哲理，形体简洁而不单调，突出公寓建筑的特征及使用的功能性要求，同时也调动建筑与环境的形态语言，共同塑造了富有生机的建筑群。建筑构思巧妙，充分利用可能的设计要素，在充分满足功能与经济实用的前提下，塑造出了生动而富有文化内涵的、具有北大人文精神及几十年留学生管理经验的经典园区，从而成为具有时代特征的北京大学又一个标志性建筑群。

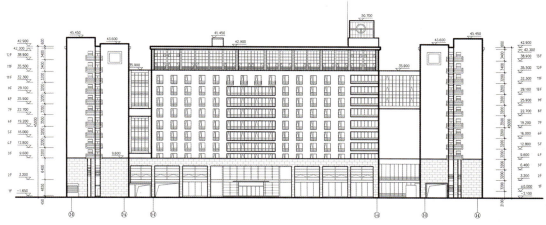

West Elevation 西立面图

KEY WORDS
关键词

Frame Structure 框架结构
Sun Louver 遮阳百叶
Empty Space 架空层

New Library of Nanjing Arts Institute
南京艺术学院新图书馆

FEATURES
项目亮点

The building is a north-south rectangular library and shares the same height with the old library. Besides, a square and flat reading space is created in the best possible way.

建筑体量是南北向长方体，底层架空以保证屋面与老馆相平，同时争取尽量方正平直的阅览空间。

Location: Nanjing, Jiangsu, China
Architectural Design: China Architecture Design & Research Group
Land Area: 6,902 m²
Floor Area: 9,956 m²
Building Height: 21.3 m
Design: 2008
Completion: 2010

项目地点：中国江苏省南京市
设计单位：中国建筑设计研究院
用地面积：6 902 m²
建筑面积：9 956 m²
建筑高度：21.3 m
设计时间：2008 年
竣工时间：2010 年

Overview

The extension project of Nanjing Arts Institute (hereinafter referred to as New Library) is on the north of the old library. It is situated in a narrow land between the student canteen and the College of Liberal Arts, which is 118 m long in south-north direction and 35 m wide in east-west direction. Since the site is on the edge of a hillside, one can see the big height difference there.

项目概况

南京艺术学院图书馆扩建工程（以下简称新馆）位于老图书馆北侧，学生食堂与人文学院之间的狭长地带，南北长约为118 m，东西最窄处仅为35 m。且恰好位于山坡的边缘，地势高差较大。

Functional Layout

Since the new library is the extension and continuation of the old library, their functions need to be coordinated and adjusted. Therefore, the reading room is arranged in the new library and office and auxiliary facilities are housed in the old one. Both of them are connected level by level that strengthens the internal circulation. Vertically speaking, the lower floors of the new library are for collecting & editing department and circulation department, while the upper space is for reading room. Down the stairs under the overhead portico is an independent art salon. And compact stacks and some school logistics rooms are arranged in a semi-underground section.

功能布局

新馆是老馆功能的拓展和延续，因此两馆的功能需要统筹调整。基本分区是新馆安排阅览室，办公等辅助功能安置在老馆内。内部交通是将新老馆各层之间等标高相接，加强了馆内功能的无缝衔接。新馆竖向分区则是底层安排图书馆的采编、流通部门，上部阅览。架空柱廊下的大台阶下还嵌入了一个功能相对独立的艺术沙龙。而密集书库及部分学校后勤用房则布置在利用场地高差的半地下部分中。

 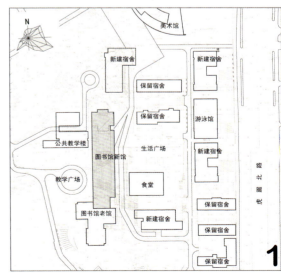

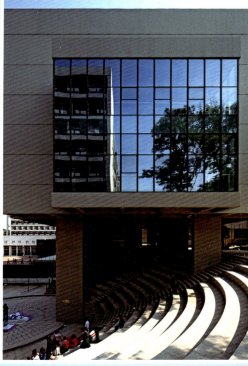

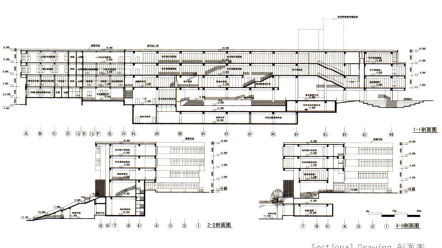
Sectional Drawing 剖面图

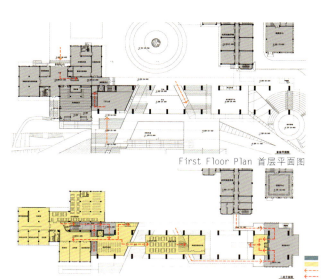
First Floor Plan 首层平面图

Second Floor Plan 二层平面图

Architectural Design

The new library is between the planning public education square and life plaza, on the only way that the students take from dormitory to teaching building. So the first two floors are emptied for a number of ramps and steps in accordance with the height difference to provide students a convenient way to the library in all directions.

Since the new library is in a crack surrounded by existing buildings, it is kept close to the surroundings as much as possible. It is a north-south rectangular one and shares the same height with the old library. Besides, a square and flat reading space is created in the best possible way. Large-span portico and east-west sun louver make the volume much more compact and avoid the feeling of clogging. Steps and ramps in the bottom space interplay with the school terrain features and realize a harmonious combination with the surrounding environment thanks to the vegetation on the slope.

建筑设计

新馆的位置处在规划的公共教育广场与生活广场之间,是学生由宿舍区到达教学区的必经之路上,因此设计将整个建筑底层架空两层,利用高差提供了大量坡道和台阶,为学生从各个方向到达图书馆以及穿越图书馆提供便捷的道路。

新馆处在周边的现状建筑的夹缝中,因此建筑形体应与周边建筑在尺度上尽量接近,建筑体量是南北向长方体,底层架空以保证屋面与老馆相平,同时争取尽量方正平直的阅览空间。大跨度的柱廊和东西向的遮阳百叶使建筑体量显得更为轻巧,避免了在周边建筑群的环抱中可能出现的堵塞感觉。架空层底部的层层台阶和坡道则与校园地势特征相应,利用坡地植被可与周围环境融为一体。

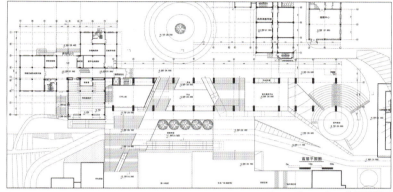
First Floor Plan 首层平面图

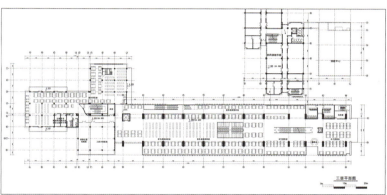
Third Floor Plan 三层平面图

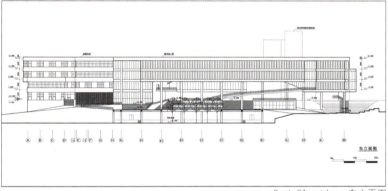
East Elevation 东立面图

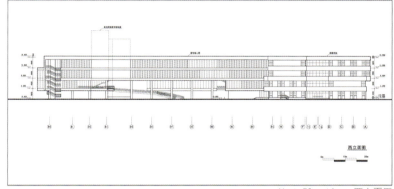
West Elevation 西立面图

KEY WORDS 关键词

Atmospheric 有美感的

Distinctive 色调鲜明

Landmark 标志性建筑

Chengde Petroleum Higher Academy Library
承德高等石油专科学院图书馆

FEATURES 项目亮点

With "applicable, flexible, efficient, economical, safe, beautiful" as the general principles, the design seeks to give full consideration of the functional convenience for teachers and students.

设计以"适用、灵活、高效、经济、安全、美观"为总体原则，在设计中力图从功能上充分考虑师生使用的便利性。

Location: China Chengde, Hebei, China
Architectural Design: Architectural Design and Research Institute of Tianjin University
Land Area: 20,585 m²
Total Floor Area: 16,466 m²
Building Height: 23.95m
Plot Ratio: 0.8

项目地点：中国河北省承德市
设计单位：天津大学建筑设计研究院
用地面积：20 585 m²
总建筑面积：16 466 m²
建筑高度：23.95 m
容积率：0.8

Overview

Chengde Petroleum Higher Academy Library is a landmark in landscape and architecture in the new campus master plan district. In addition to the basic function as an information center and public space, the design strives to meet the current requirements as well as for future upgrades through building orientation and morphology of scrutiny.

项目概况

承德高等石油专科学院图书馆是新校区总体规划中心区的重要景观和标志性建筑。单体设计力求在体现图书馆建筑作为信息中心及公共活动场所应具有特质的基础上，通过建筑定位及形态的推敲，力争将其打造成既符合当下使用要求，同时又能进行未来升级改造的新型图书馆建筑。

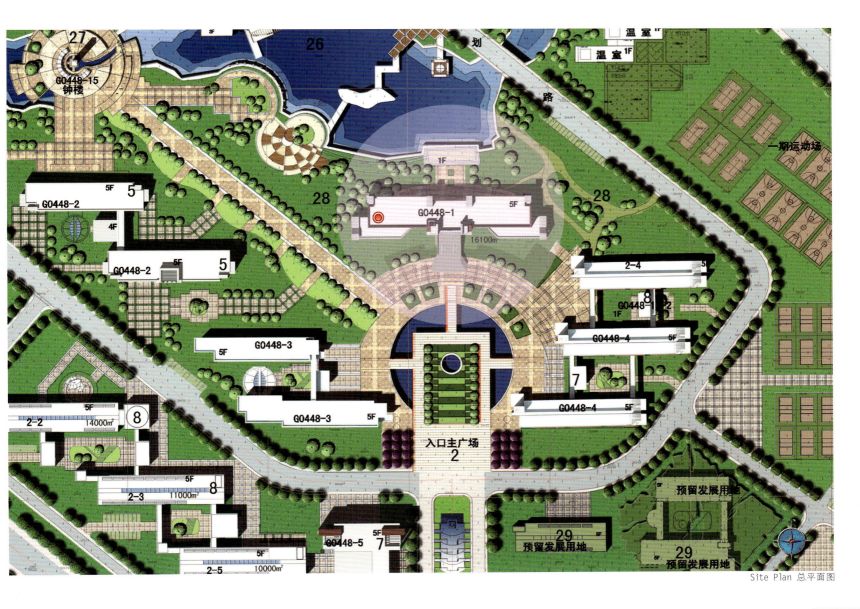

Site Plan 总平面图

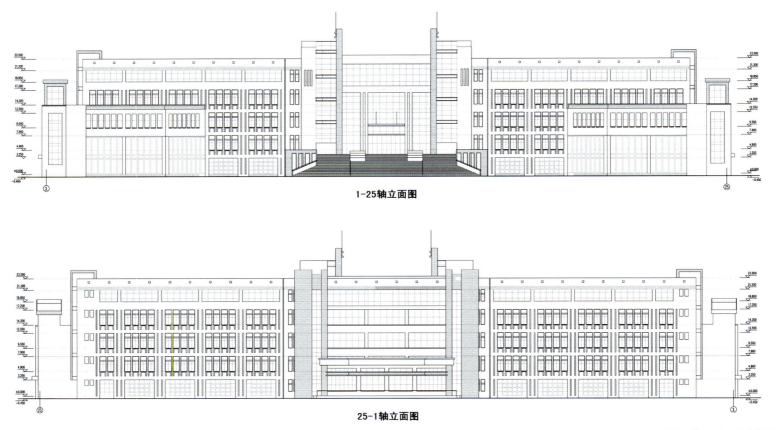

1-25轴立面图

25-1轴立面图

Side Elevation 轴立面图

Layout

A total of five floors, the first floor houses school history exhibition hall, editing and intensive library room, lecture hall of 300 people and the electronic reading room. The second floor houses the entrance hall, the retrieval and circulation office and reading room. The third floor is used for reading and lending books in chinese and foreign languages. The fourth floor serves as chinese and foreign periodicals reading and lending area without separate lounge, there are seats for a rest or study in corridors and some reading rooms or study hall. The fifth floor houses activity rooms, classrooms, network center and information center.

平面布局

建筑共五层，首层设校史陈列厅、采编及密集书库、300人学术报告厅、电子阅览室等。二层为入口大厅、检索、借还处、阅览室。三层为中外文图书藏借阅览室。四层为中外文期刊藏阅借阅区。不另设休息室，但回廊及部分阅览室可设座椅作为休息或自习座位。五层设活动室、教室、网络中心及信息中心。

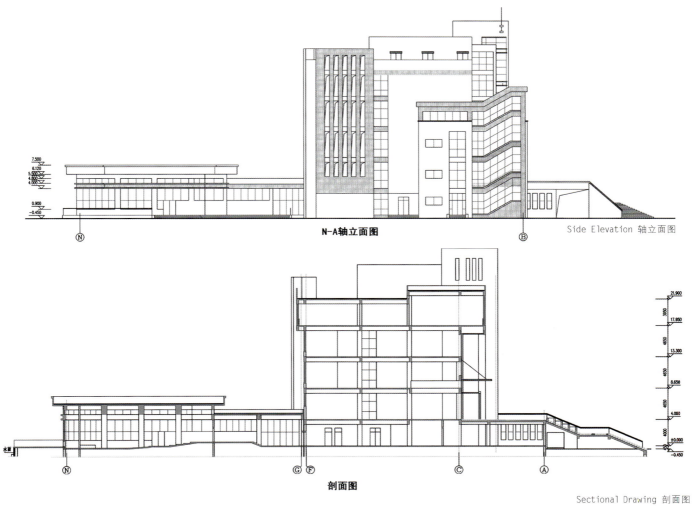

N-A轴立面图　　Side Elevation 轴立面图

剖面图　　Sectional Drawing 剖面图

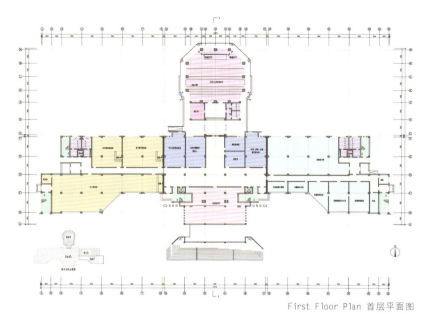
First Floor Plan 首层平面图

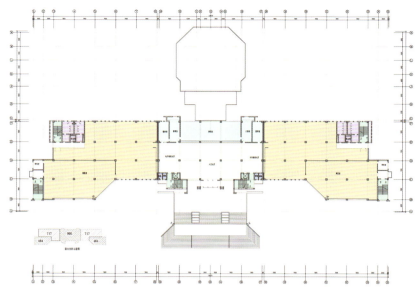
Second Floor Plan 二层平面图

Design concept

With "applicable, flexible, efficient, economical, safe, beautiful" as the general principles, the design seeks to give full consideration of the functional convenience for teachers and students. It also considers the decentralized and centralized circulation modes to future changes in management mode and there is a decentralized reading and lending area in each reading room. The ideal Layout for reading and lending books, both for teachers and students and both for books and editorials; the design creates a good environment and working conditions for students and staff.

设计理念

设计以"适用、灵活、高效、经济、安全、美观"为总体原则，在设计中力图从功能上充分考虑师生使用的便利性，设计时考虑分散借还和集中借还两种模式，以便今后随形势改变选择管理模式，在各阅览室设有分散借阅处。达到借阅合一、师生合一、书刊合一的理想布局方式，为师生和工作人员创造良好的借阅环境和工作条件。

KEY WORDS
关键词

Tortuous Space 曲折空间

Suzhou Features 苏州特色

Humanistic Implication 人文意蕴

Dushuhu Higher Education District Southeast University Suzhou Graduate School Phase II

独墅湖高教区东南大学苏州研究生院二期工程

FEATURES
项目亮点

The design gives priority to the white and also goes with moderate gray building facades, which remind people of Suzhou ancient city characteristics of white wall tiles; in addition, it establishes a "winding" "sequence" to express the quiet and deep cultural connotation of ancient Suzhou, "curved rather than straight", "winding rather than obstructed".

设计采用以白色为主，搭配适度灰色的建筑外墙，使人联想到粉墙黛瓦的古城苏州特色，并且设计通过建立一个"曲折"的"秩序"，进行微妙的显露来表达古城苏州"曲而不直"、"折而不通"、静谧而幽深的人文意蕴。

Location: Suzhou, Jiangsu, China
Architectural Design: Architects & Engineers Co., Ltd. of Southeast University
Total Land Area: 66,770 m²
Total Floor Area: 21,778 m²
Plot Ratio: 0.95

项目地点：中国江苏省苏州市
设计单位：东南大学建筑设计研究院有限公司
总用地面积：66 700 m²
总建筑面积：21 788 m²
容积率：0.95

Overview

The project is a group of modernist style education buildings to express the meaning of Suzhou specialty regionalism. Planning and architectural design attempts to convey three basic concepts to the public: modern, Suzhou, Southeast University, thus forming a group of functional and practical, and reasonable buildings with modest cost. The project will be a space with formal beauty without exaggeration. The Institute buildings have a total of eight blocks, no. one to five buildings for the first period and No. six, seven, eight buildings for the second period which are mainly for the party office and research laboratory.

项目概况

项目是一组以现代主义风格来表达苏州特色地域主义内涵的教育建筑群。规划与建筑设计试图向公众传达三个基本概念：现代的、苏州的、东大的，从而形成一个功能实用、造价适度、建造合理、空间与形式美观而不虚夸的空间场所。研究院总共由八幢建筑组成，其中一号楼到五号楼为一期工程，六、七、八号楼为二期工程，主要为党政办公楼与研发实验楼。

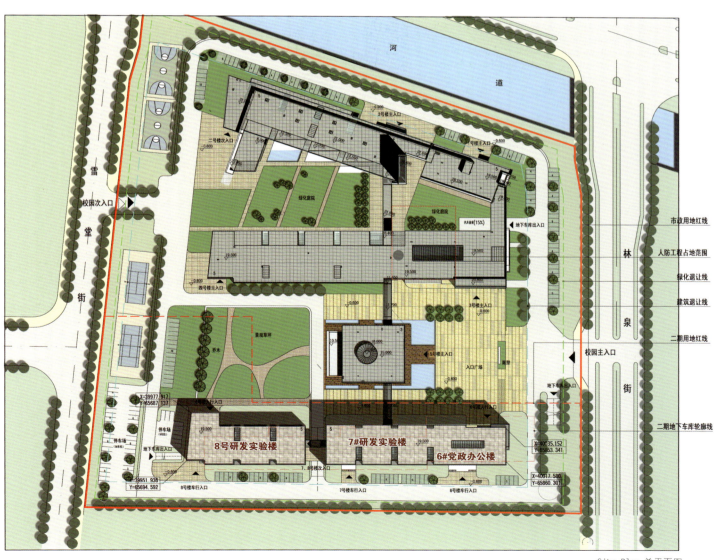

Site Plan 总平面图

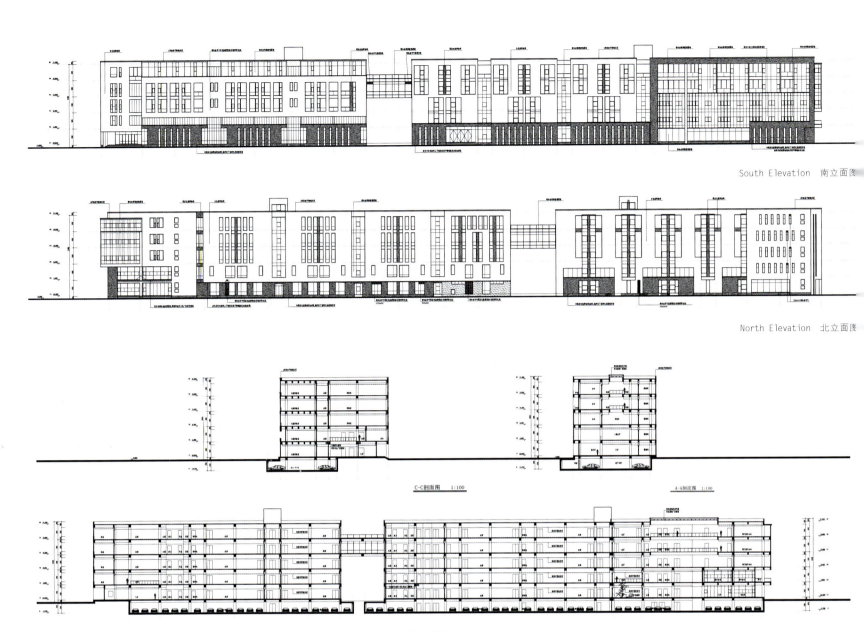

South Elevation 南立面图

North Elevation 北立面图

C-C剖面图 1:100　　A-A剖面图 1:100

B-B剖面图 1:100

Sectional Drawing 剖面图

Planning Layout

By using the concept a "making the city" to "create a compact, land closely and accommodate teaching, research and the city's education office in different functional areas", the planning provides an integrated urban spatial structure: a miniature "compact city" with a horizontally continuous multi-scale free extension to enhance the exchange interaction between teachers and students.

规划布局

规划总体上采用"造城"理念——"造一座紧凑、节地、联系紧密而容纳教学、科研与办公不同功能区域的教育之城",从而提供一个整合的城市空间架构：以一个水平向自由伸展的连续的多层尺度形体形成一个微缩的"紧凑城市",所形成的紧凑的交流空间可加强师生间的互动交流。

Facade design

The design gives priority to the white and also goes with moderate gray building facades, which remind people of Suzhou ancient city characteristics of white wall tiles; in addition, it establishes a "winding" "sequence" to express the quiet and deep cultural connotation of ancient Suzhou, "curved rather than straight", "winding rather than obstructed". This is mainly for breaking the system of orthogonal axis in the overall network layout and instead twisting and folding it to produce a winding space; the building body is the dislocation and extension of primitive geometry of Plato in order to achieve continuity of space; the project facade adopts more "folding" rather than straight lines in order to achieve the echo of form.

建筑立面设计

设计采用白色为主,搭配适度灰色的建筑外墙,使人联想到粉墙黛瓦的古城苏州特色,并且设计避免采用直白的西方建筑语言的模式,而是通过建立一个"曲折"的"秩序",进行微妙的显露来表达古城苏州"曲而直"、"折而不通"、静谧而幽深的人文意蕴。这主要表现为在总体布局上打破正交的轴网体系而进行扭转和折叠,从而产生曲折的空间；建筑形体打破柏拉图体的几何原型而进行错动和延伸,以达到空间的延续；在建筑立面上采用更多的"折线"而非直线以求得形式的呼应。

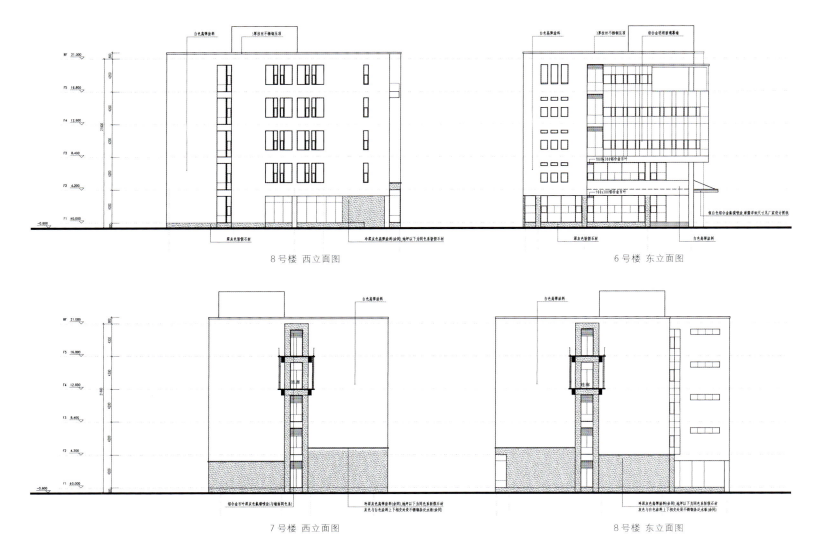

8号楼 西立面图　　　　　　　　　　6号楼 东立面图

7号楼 西立面图　　　　　　　　　　8号楼 东立面图

Elevation 立面图

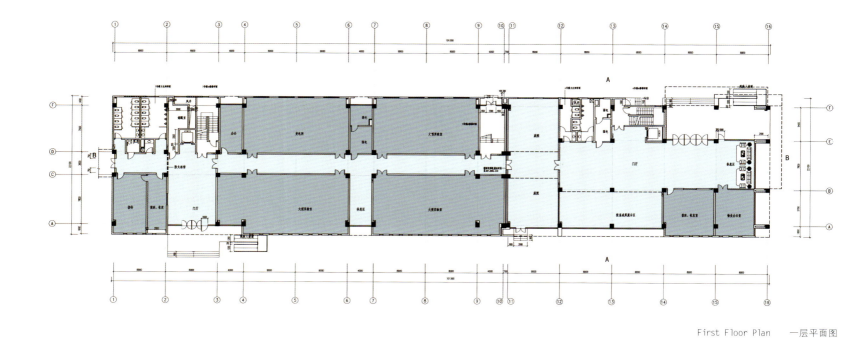

First Floor Plan　一层平面图

KEY WORDS 关键词	Arc-shaped Appearance 弧形外观
	Glass Facade 玻璃立面
	Multiple Functions 功能齐全

Arts Centre of Shandong University of Technology
山东理工大学大学生艺术中心

FEATURES 项目亮点

The building encloses the sunken square to highlight the central role of "Green Island" central axis; it echoes the library in southern side at a distance, and acts as transition, convergent or connection from teaching area to dormitory area.

该建筑通过围合下沉广场，加强学校"绿岛"中轴线的中心作用，与南面的图书馆遥相呼应，并起到了自教学区到学生宿舍区的"过渡、收口、连接"作用。

Location: Zibo, Shandong, China
Architectural Design: Shandong Tongyuan Group Corporation
Area of Base: 6,900.0 m²
Site Area: 26,650.0 m²
Total Floor Area: 15,918.0 m²
Plot Ratio: 0.6
Green Area Ratio: 60%

项目地点：中国山东省淄博市
设计单位：山东同圆设计集团有限公司
基底面积：6 900.0 m²
用地面积：26 650.0 m²
总建筑面积：15 918.0 m²
容积率：0.6
绿地率：60%

Planning and Layout

The building encloses the sunken square to highlight the central role of "Green Island" central axis; it echoes the library in southern side at a distance, and acts as transition, convergent or connection from teaching area to dormitory area. This project is comprehensive complex that contains service rooms and offices, and provides students with public gathering space and performing space, including a 1,700-seats large multi-purpose halls, which provides an important platform for school service, work, communication and performance.

规划布局

该建筑通过围合下沉广场，加强学校"绿岛"中轴线的中心作用，与南面的图书馆遥相呼应，并起到了自教学区到学生宿舍区的"过渡、收口、连接"作用。工程为内含学生服务及办公并为学生提供公共集会和表演的复杂综合体，其中包含1 700座的大型多用途厅，是提供学生与学校服务、办公、交流、会演的重要平台。

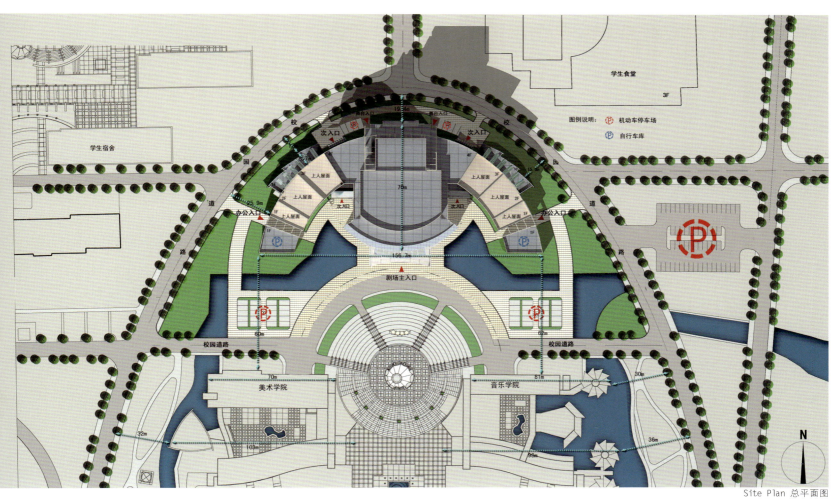

Site Plan 总平面图

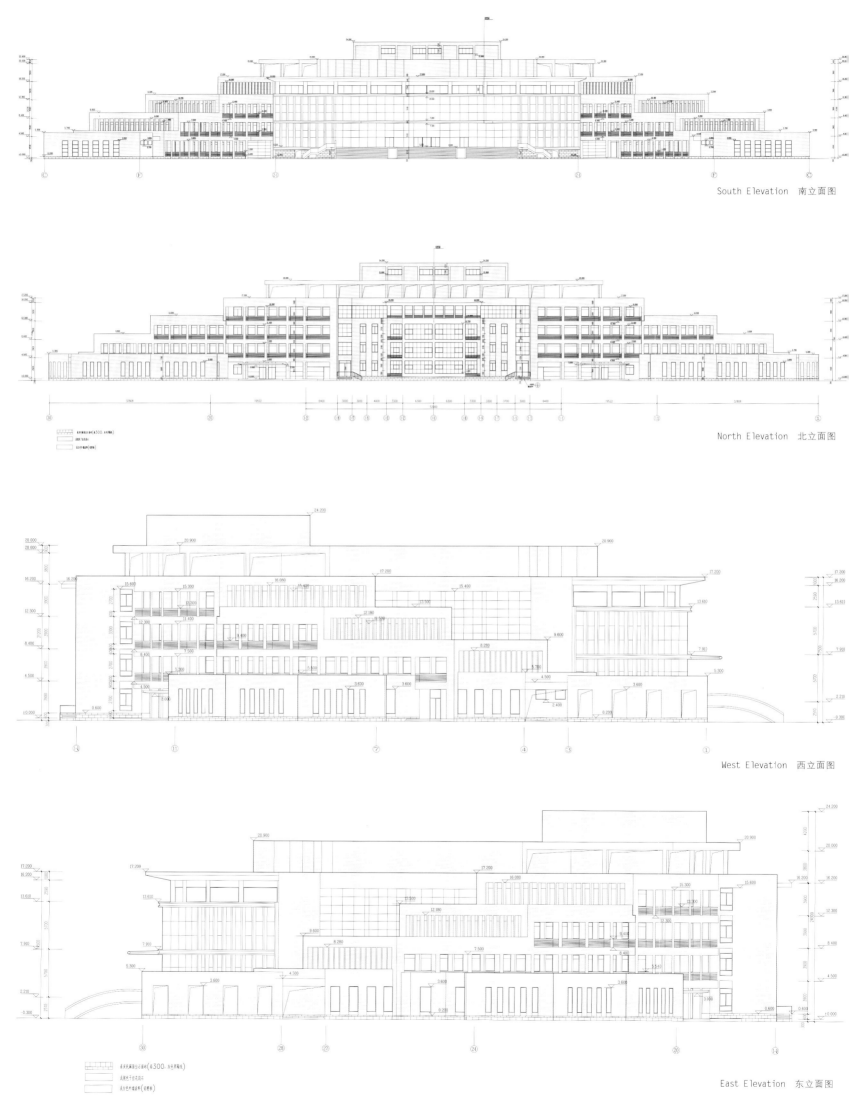

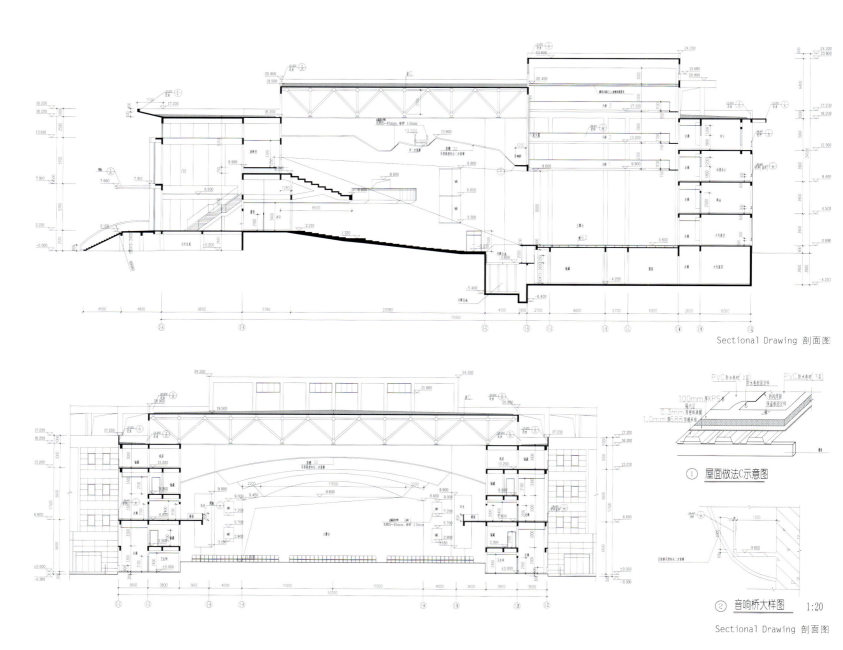

Sectional Drawing 剖面图

① 屋面做法C示意图

② 音响桥大样图 1:20

Sectional Drawing 剖面图

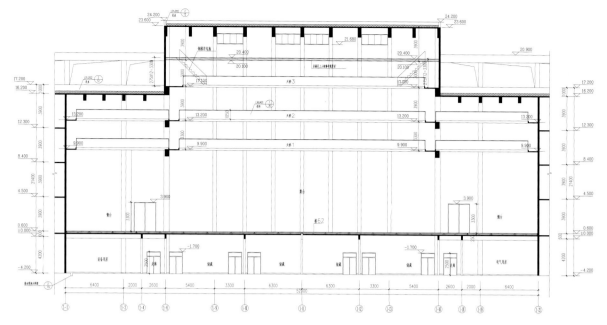

Sectional Drawing 剖面图

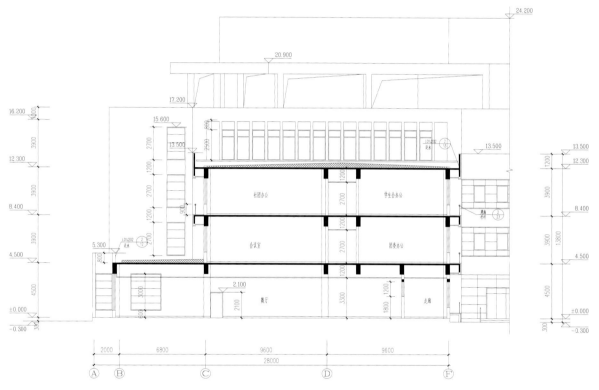

Sectional Drawing 剖面图

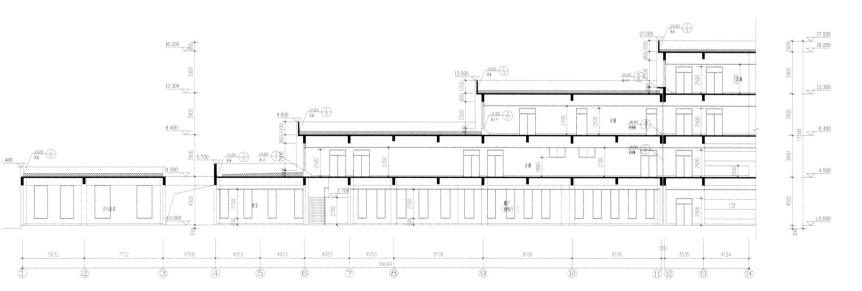

Sectional Drawing 剖面图

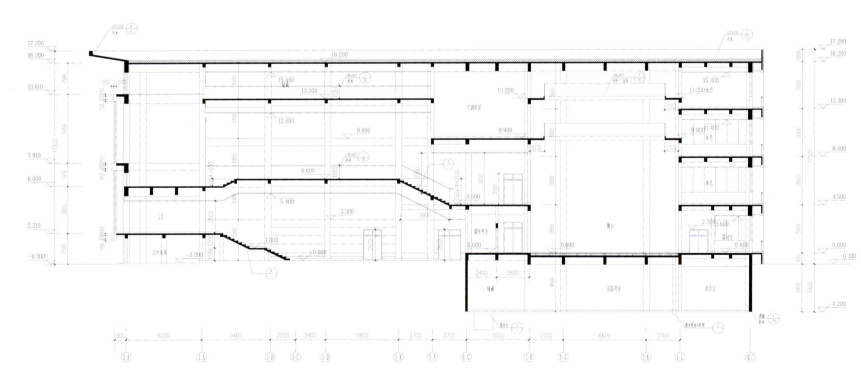

Sectional Drawing 剖面图

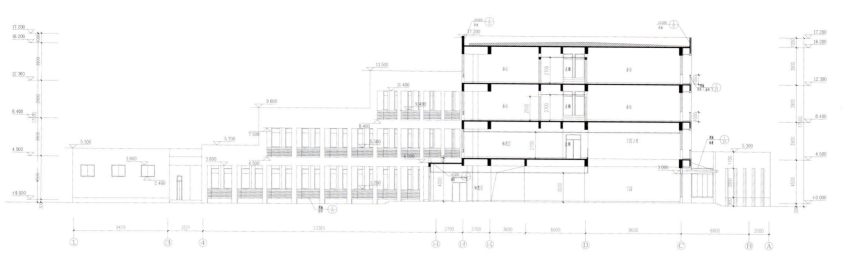

Sectional Drawing 剖面图

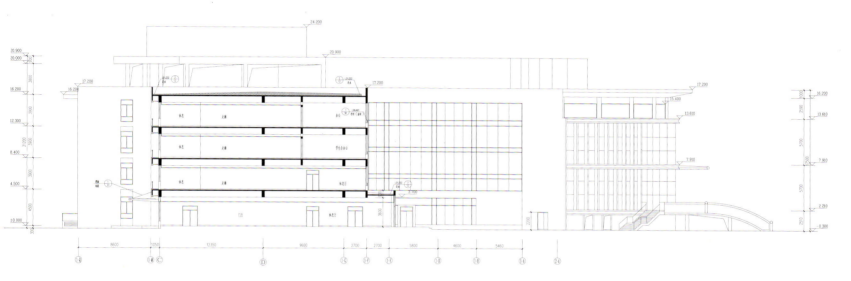

Sectional Drawing 剖面图

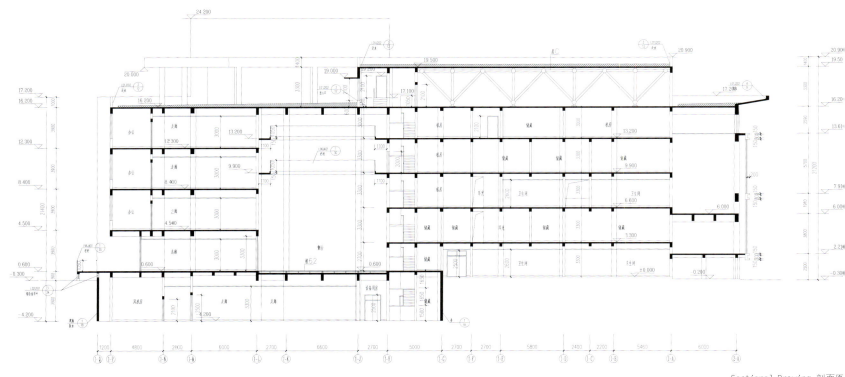
Sectional Drawing 剖面图

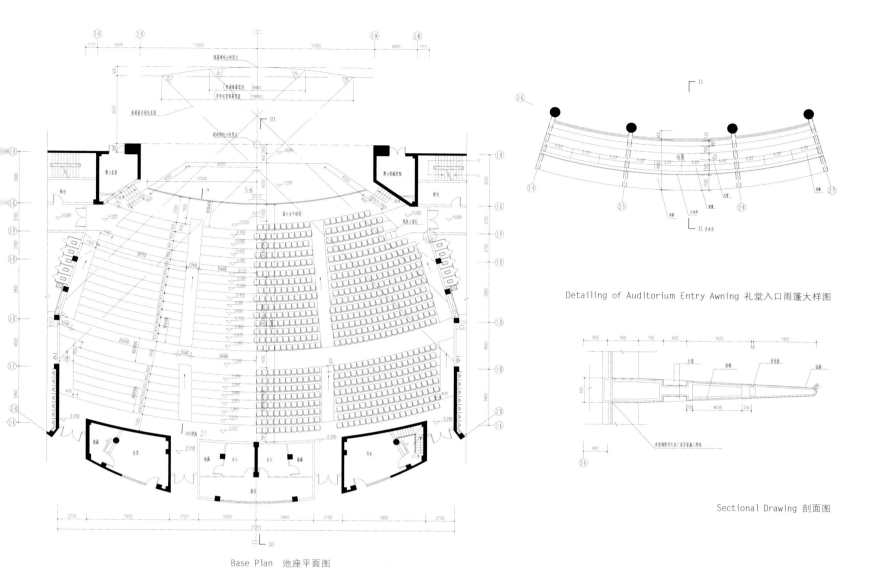

Detailing of Auditorium Entry Awning 礼堂入口雨篷大样图

Sectional Drawing 剖面图

Base Plan 池座平面图

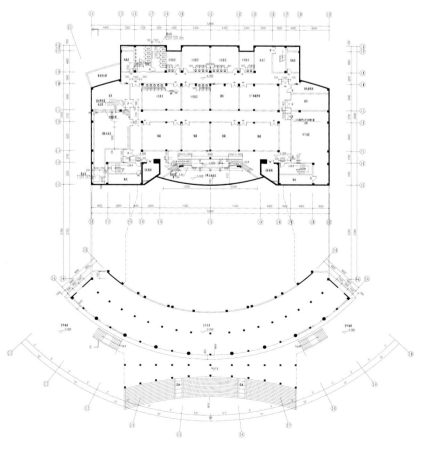

Plan for Basement Floor 地下一层平面图

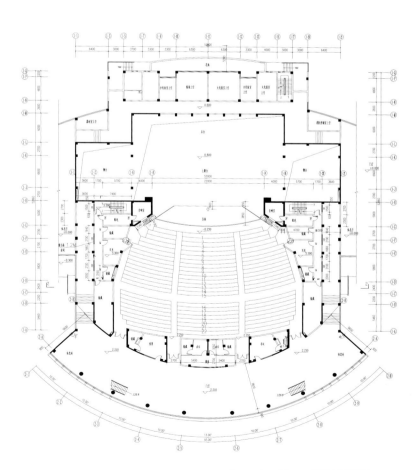

Mezzanine Plan 一层夹层平面图

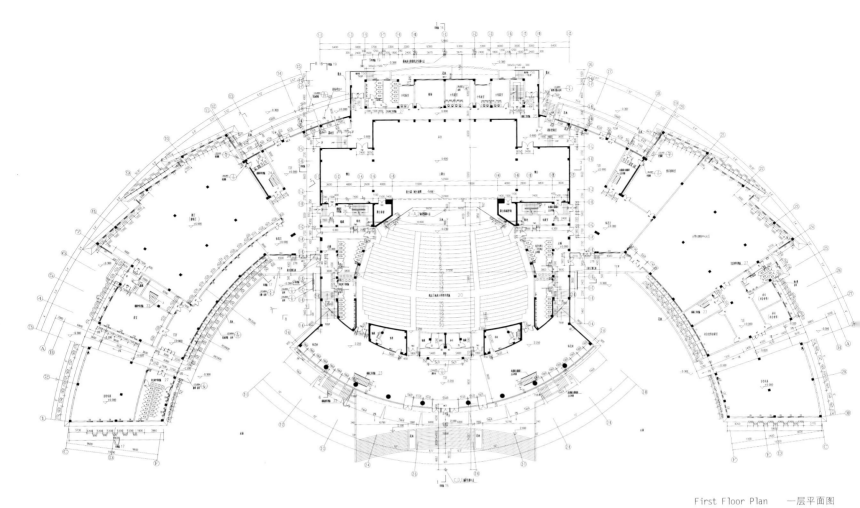

First Floor Plan 一层平面图

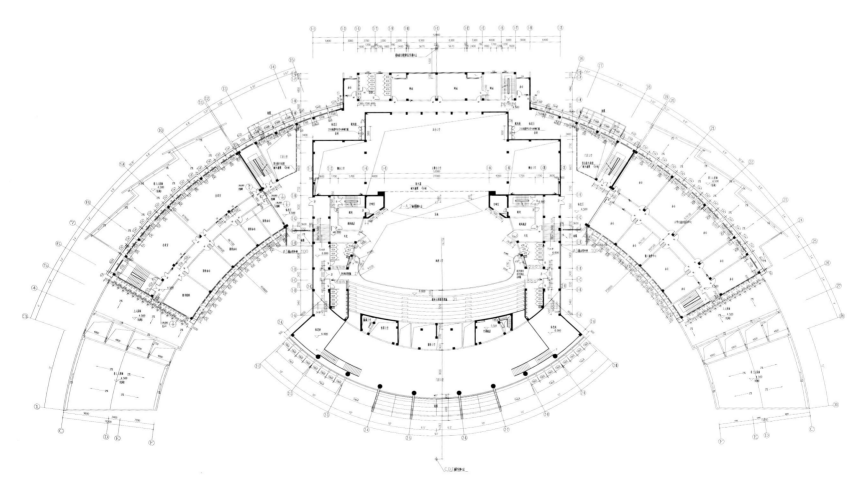

Second Floor Plan 二层平面图

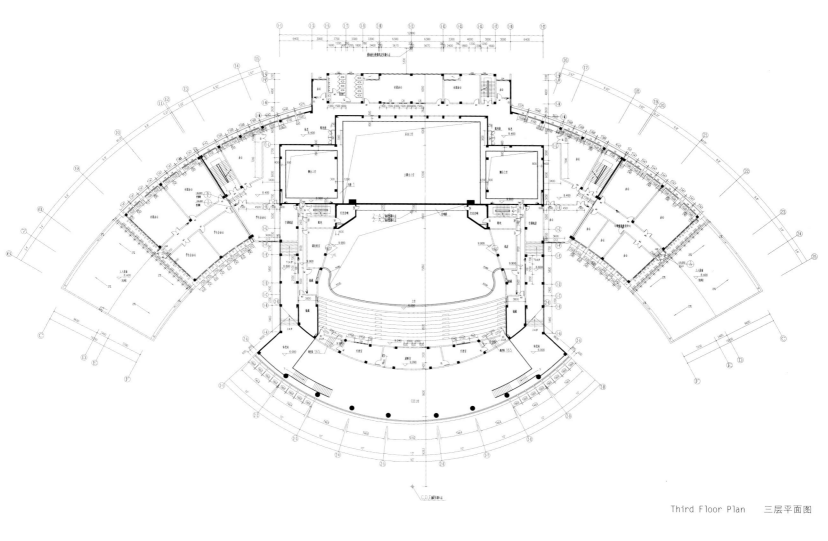

Third Floor Plan　三层平面图

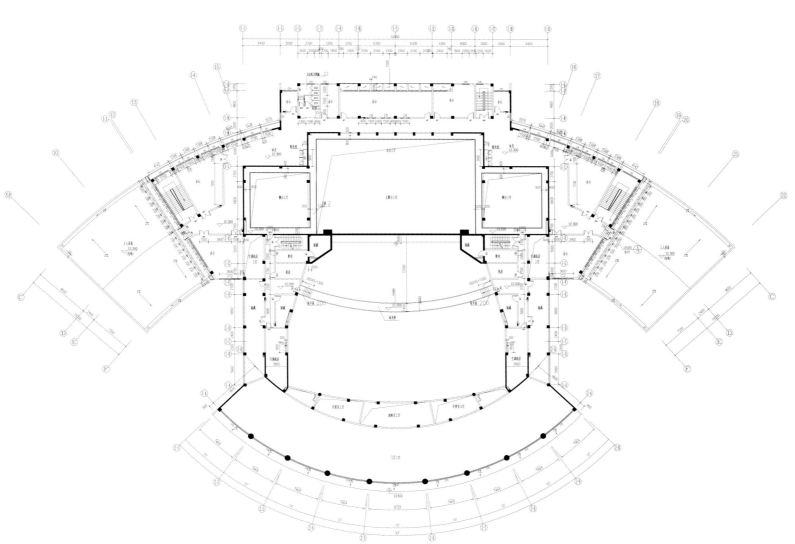

Fourth Floor Plan　四层平面图

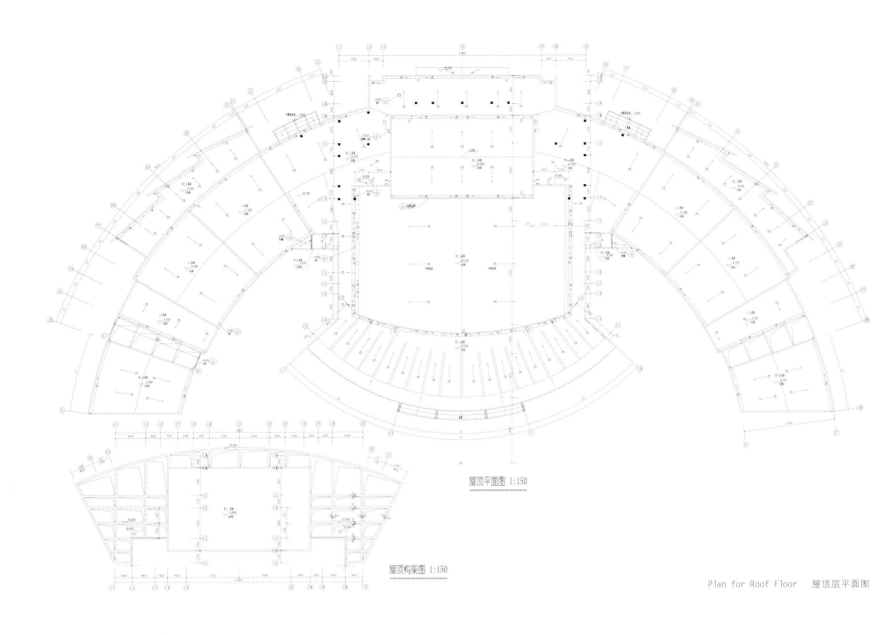

屋顶平面图 1:150

屋顶构架图 1:150

Plan for Roof Floor　屋顶层平面图

KEY WORDS 关键词

Fan-shaped Plane 扇形平面
Arc-shaped Surface 弧形界面
Spectacular Appearance 外观醒目

Library of Shandong University of Technology
山东理工大学图书馆

FEATURES 项目亮点

The northern side of the library presents as a surrounded arc and develops into a square with proper scale, makes the shape of the library look likes opened arms, which integrates with the strong axis space sequence of northern teaching area.

图书馆的北侧呈内向环抱的弧形，形成了一个尺度适宜的广场，使图书馆建筑宛如伸开的双臂，接纳了北侧教学区强烈的轴线型空间序列。

Location: Zibo, Shandong, China
Site Area: 17,082 m²
Total Floor Area: 35,479.6 m²
Floor Area aboveground: 27,749.2 m²
Floor Area underground: 7,730.4 m²
Greenland Area: 3,775 m²
Building Density: 53.13%
Plot Ratio: 1.64
Greening Rate: 22.1%

项目地点：中国山东省淄博市
总用地面积：17 082 m²
总建筑面积：35 479.6 m²
地上建筑面积：27 749.2 m²
地下建筑面积：7 730.4 m²
绿地面积：3 775 m²
建筑密度：53.13%
容积率：1.64
绿化率：22.1%

Architectural Design

Library of Shandong University of Technology is located in the northern side of the oval green axis, arc surface applied in the southern side finally develops green axis, and it also echoes the architectural form in the northern side of the green axis. The northern side of the library presents as a surrounded arc and develops into a square with proper scale, makes the shape of the library look like opened arms, which integrates with the strong axis space sequence of northern teaching area. The overall fan-shaped plane is divided into four sectors by three vertical axes. The lift well and other important public service facilities are set in the vertical axes area, they interwine with lateral arc-shaped alleyway, and constitute the internal traffic network in this library.

建筑设计

山东理工大学图书馆建筑位于椭圆绿轴的南端，南侧采用弧形界面是对绿轴的最终完形，也是对绿轴北端建筑形态的呼应。图书馆的北侧呈内向环抱的弧形，形成了一个尺度适宜的广场，使图书馆建筑宛如伸开的双臂，接纳了北侧教学区强烈的轴线型空间序列。图书馆的总体扇形平面又被三条纵轴切分为四个扇形。纵轴内包含楼电梯间等重要的公共服务设施，与横向的弧形通道交织在一起，构成了图书馆内部的交通网络。

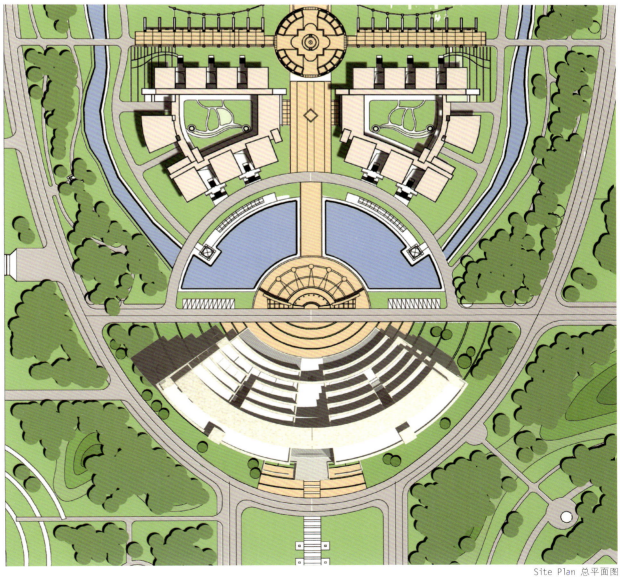

Site Plan 总平面图

South Elevation 南立面图

North Elevation 北立面图

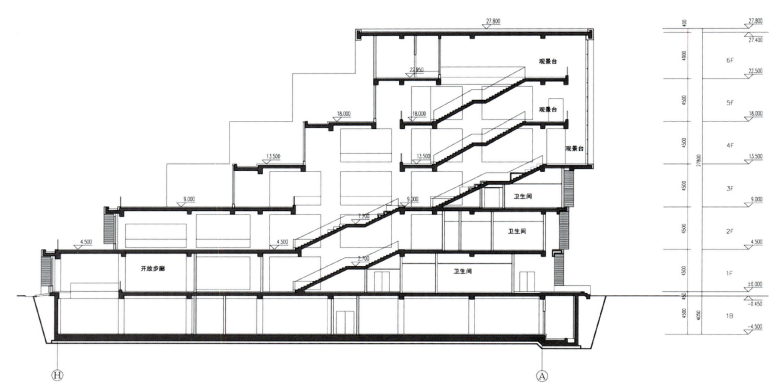

Sectional Drawing 剖面图

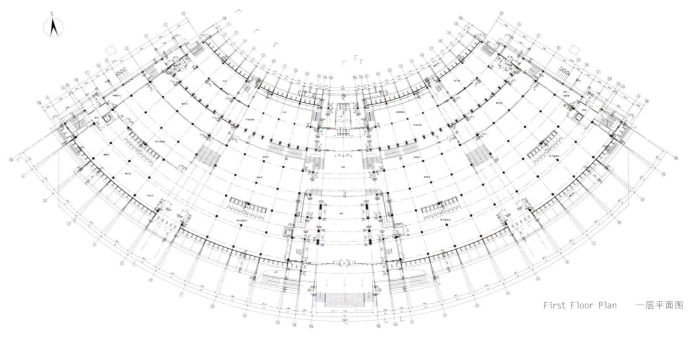

First Floor Plan　一层平面图

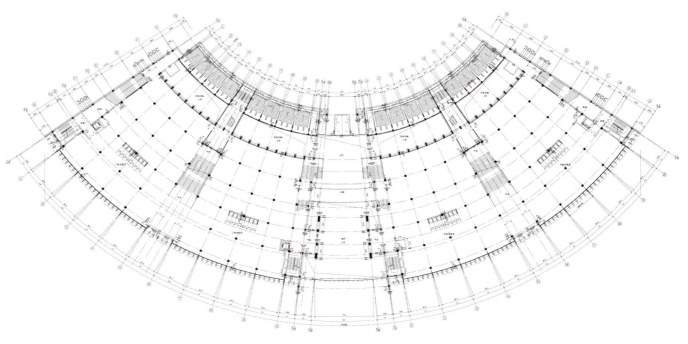

Second Floor Plan　二层平面图

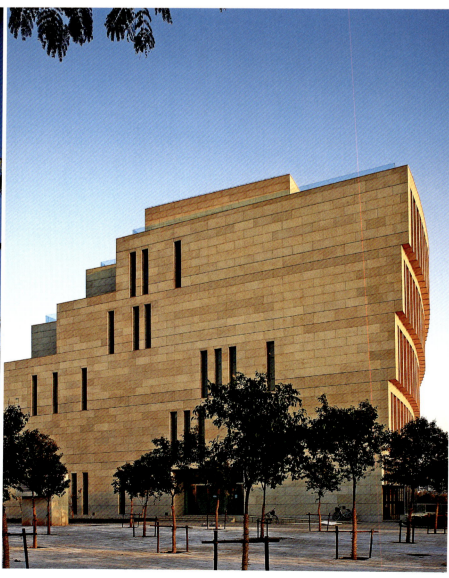

KEY WORDS 关键词

Simple Appearance 外观简洁

Integrated Environment 融于环境

Unique 独具特色

Fujian South China Women's Vocational College New Campus
福建华南女子职业学院大学城新校区

FEATURES 项目亮点

The design carefully handles mass relationship according to layout and its location. Gentle and simple hip roof, wall details consistent with slab column, windows and other functional components which have a proper ratio of the scale and structure don't employ excessive whitewash to reach a clean, modern, simple appearance image.

项目结合单体平面布局及所处区位精心处理体块关系，屋面以平缓简约的四坡为主，墙身细部结合梁板柱窗等功能构件的比例尺度与构造关系，不做多余的粉饰而达成简洁、现代、朴实的外观形象。

Location: Fuzhou, Fujian, China
Architectural Design: Architectural Design and Research Institute of Fujian Province
Land Area: 101,653 m²
Total Building Area: 71,820 m²
Building Base Area: 18,700 m²
Building Density: 16.7%
Plot Ratio: 0.643
Green Rate: 38.4%

项目地点：中国福建省福州市
设计单位：福建省建筑设计研究院
用地面积：101 653 m²
总建筑面积：71 820 m²
建筑基底面积：18 700 m²
建筑密度：16.7%
容积率：0.643
绿地率：38.4%

Architectural Design

Under the campus master plan, by making full use of the old campus building art symbol, the single building employs exterior prominent to emphasize women college's temperament and characteristics and integrates with the surrounding campus natural landscape. The design carefully handles mass relationship according to layout and location. Gentle and simple hipped roof, wall details consistent with slab column, windows and other functional components which have a proper ratio of the scale and structure don't employ excessive whitewash to reach clean, modern, simple appearance image. Using blue-gray tile roof, exterior red brick and white upper wall paint, wall paneling made of cyan rubble, colorless, transparent and white glass and dark gray railings, window, Alice eaves, the building forms both the old south China style, harmony with the surrounding architecture, and unique architectural image.

建筑设计

单体建筑外观充分运用老华南建筑群建筑艺术符号，突出女性高等院校的气质和特色，又根据大学城总体规划，与周边已建成校园及自然景观协调、融合。项目结合单体平面布局及所处区位精心处理体块关系，屋面以平缓简约的四坡为主，墙身细部结合梁板柱窗等功能构件的比例尺度与构造关系，不做多余的粉饰而达成简洁、现代、朴实的外观形象。建筑采用蓝灰瓦屋面，外墙上部为红色面砖与白色外墙涂料，外墙墙裙为青色毛石，无色透明白玻璃及深灰色栏杆、窗饰、翘屋檐，形成既有老华南的建筑风格且与周边建筑协调，又独具品位的建筑形象。

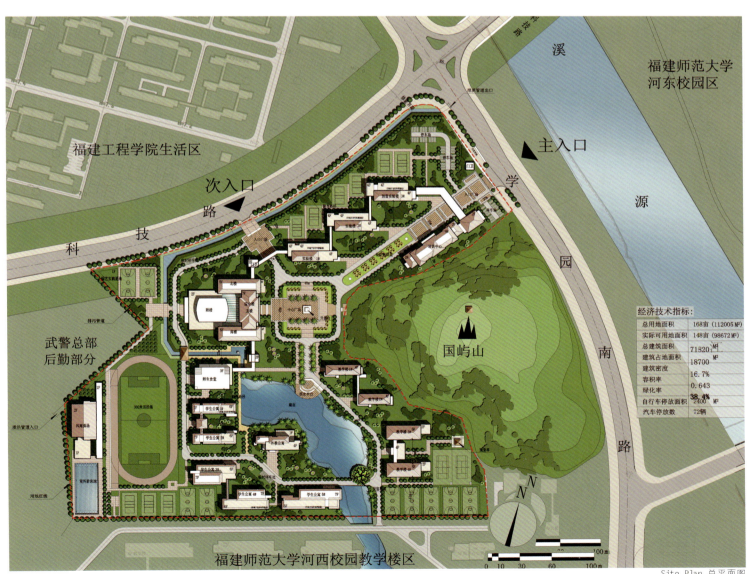

Site Plan 总平面图

①～⑮立面图

⑰～Ⓐ立面图

Elevation 立面图

Sectional Drawing 剖面图

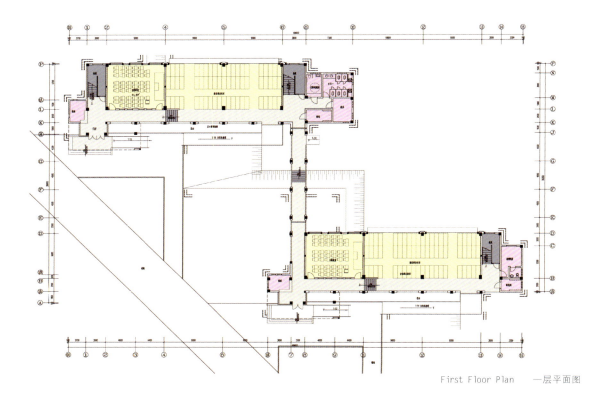

First Floor Plan　一层平面图

KEY WORDS 关键词

Beautiful Appearance 外观优美

Simple Cube Shape 形体简洁

Body Mass 方体体量

Yifu Library, West Lake Campus of Fuyang Normal College
阜阳师范学院西湖校区逸夫图书馆综合楼

FEATURES 项目亮点

As an important campus new building, the project follows the architectural design principles as "modern, practical, economic, aesthetic", using basic physical elements — cube and organic connection — to express the idea of the designer.

项目遵循"现代、实用、经济、美观"的建筑设计原则,作为新校区的重要建筑,采用基本的形体元素——立方体及有机的连接——来表达设计师的设计构思。

Location: Fuyang City, Anhui Province, China
Architectural Design: Architectural Design & Research Institute of Hefei Technology University
Land Area: 33,000 m²
Total Building Area: 43,910 m²
Ground Floor Area: 41,760 m²
Underground Construction Area: 2,150 m²
Plot Ratio: 1.33
Greening Rate: 30%

项目地点:中国安徽省阜阳市
设计单位:合肥工业大学建筑设计研究院
用地面积:33 000 m²
总建筑面积:43 910 m²
地上建筑面积:41 760 m²
地下建筑面积:2 150 m²
容积率:1.33
绿化率:30%

Overview

Located in the central axis of West Lake Campus of Fuyang Normal College, Yifu Library has a collection of 1 million books and functions as office, teaching, library, auditorium, etc.

项目概况

阜阳师范学院西湖校区逸夫图书馆综合楼位于阜阳师范学院西湖校区中轴线上。功能为办公,教学,图书馆,礼堂等。图书馆藏书100万册。

Design Principles

As an important campus new building, the project follows the architectural design principles as "modern, practical, economic, aesthetic", using basic physical elements — cube and organic connection— to express the idea of the designer. A variety of functions are integrated into two contrasting bodies—the five-layer lateral mass and 15-layer vertical mass —which are connected through the hall to make the project unified.

设计原则

项目遵循"现代、实用、经济、美观"的建筑设计原则,作为新校区的重要建筑,采用基本的形体元素——立方体及有机的连接——来表达设计师的设计构思。将众多功能整合在两个形成对比的体量中,即15层的竖向体块和5层的横向体块,两者通过礼堂联系使之成为统一的整体。

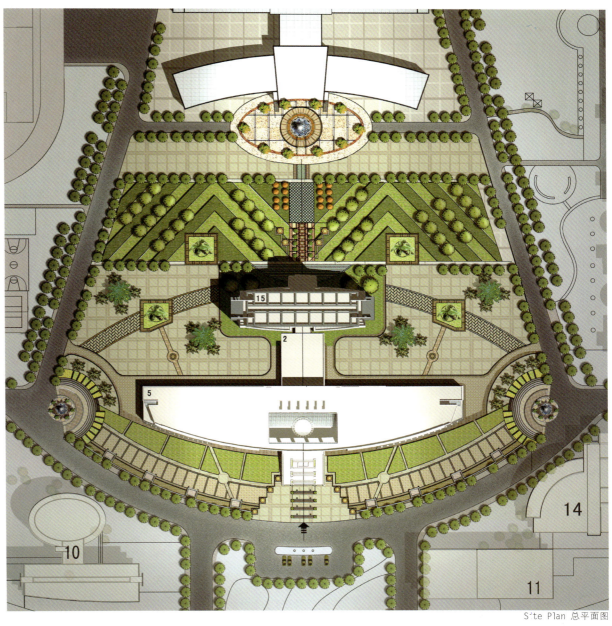

Site Plan 总平面图

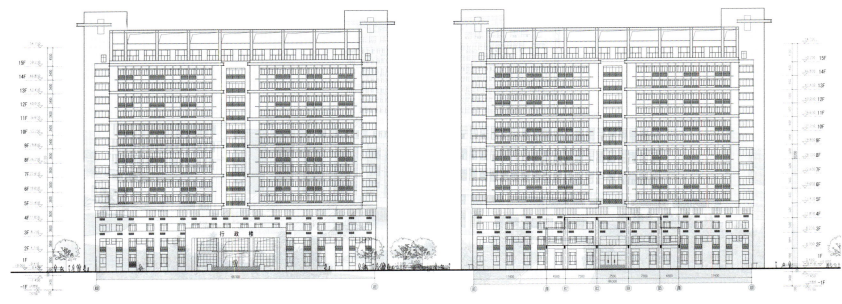
Sectional Drawing 剖面图

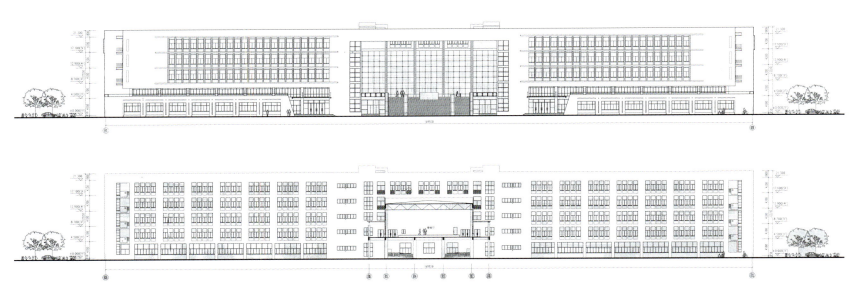
Elevation 立面图

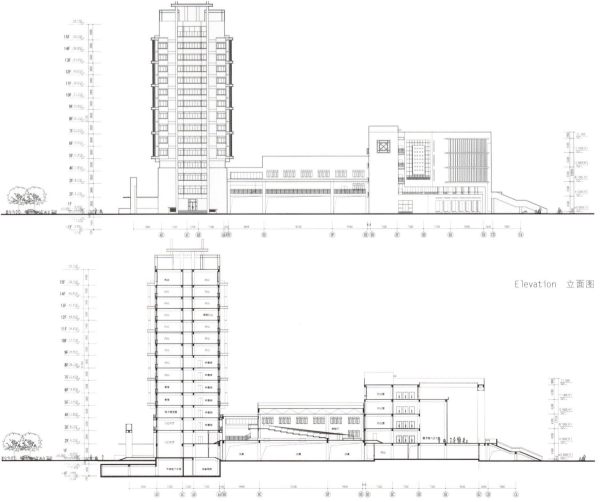

Elevation 立面图

Sectional Drawing 剖面图

Plan for Basement Floor 地下一层平面图

First Floor Plan 一层平面图

Architectural Design

Laterally expanding in space, the building looks beautiful. It looks both complete and unified from the outside, reflecting the beat and rhythm and its quiet and intimate inside. The outside vertical column of 600 mm × 1,500 mm formed an uninterrupted horizontal plate surface. Space bar has a strong sense of perspective, and the plate corresponds to the wall, there are sheet walls of a variety of spatial scales ; parallel paths guide space; the wall contains "lead ", " turn", "surround " multiple meanings. Clearance between walls and windows which form a landscape frames for walls, provide a wealth of sight angles.

建筑设计

空间上沿横向展开，建筑物舒展优美。从外看完整统一，体现了节奏和韵律感；在内部有安静温馨的场所感。外敞廊竖向由 600 mm×1500 mm 的间隔柱组成，形成不间断的水平板面。直板空间有强烈的透视感，与板对应的是墙，空间中活跃着各种尺度的片墙，平行路径引导空间走向，墙包含了"引"、"转"、"围"的多重含义。墙与墙的间隙，开窗的方式，形成了墙的景框，提供了丰富的视线角度。

KEY WORDS 关键词

"L" Shaped Layout　"L"形布局

Simple Shape　形体简洁

Elegant Environment　环境优美

Sichuan Branch of the National Prosecutors College
国家检察官学院四川分院

FEATURES 项目亮点

According to the specific circumstances of the land and combined with functional requirements, the design breaks the traditional academic layout and uses "L" type which combines the function of each organic part.

设计打破了传统学院的行列式布局，根据用地的具体情况，结合功能需求，总体采用"L"形建筑布局，将各部分使用功能有机结合在一起。

Location: Chengdu, Sichuan, China
Architectural Design: Architectural Design and Research Institute of Tsinghua University
Total Land Area: 72,100 m²
GFA: approximately 30,000 m²
Plot Ratio: 0.4

项目地点：中国四川省成都市
设计单位：清华大学建筑设计研究院有限公司
总用地面积：72 100 m²
总建筑面积：约 30 000 m²
容积率：0.4

Overview

The site is a land of 180 m wide (east-west), 360 m long (north-south). The project includes training and teaching buildings, conferences, student apartments, fitness activities, restaurants, underground garage, staff quarters and other ancillary buildings. The main building has 11 floors aboveground, 1 basement, with a total height of 44.4 m.

项目概况

项目基地是一块东西宽约 180 m，南北长约 360 m 的长方形用地。项目包括培训教学、会议、学员公寓、健身活动、餐厅等功能用房以及地下车库、职工宿舍等配套用房。项目主楼地上 11 层，地下 1 层，总高度为 44.4 m。

Layout

According to the specific circumstances of the land and combined with functional requirements, the design breaks the traditional academic layout and uses "L" shape which combines the functions of organic parts.

The collage gate faces the Lishan Boulevard in the south, and the hotel gate is in the west side facing the city road to avoid interference to the teaching building. The sports activities in central area and dining area serve as a linkage connecting with the teaching buildings to achieve the integration of accommodation, dining, and teaching.

Site Plan 总平面图

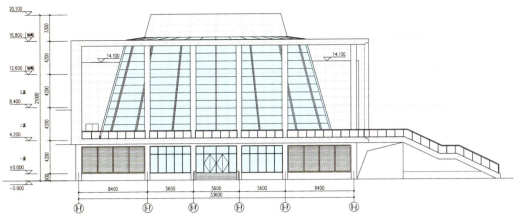

Side Elevation 轴立面图

Side Elevation 轴立面图

Meanwhile, combining with green mound natural environment in the east, the design sets green landscape, playground and green parking to create an elegant environment while reducing interference to the building. By maintaining the existing north ponds to be transformed, the design forms open space available for students to walk and talk and a beautiful lake scenic area.

规划布局

设计打破了传统学院的行列式布局，根据用地的具体情况，结合功能需求，总体采用"L"形建筑布局，将各部分使用功能有机结合在一起。

学院对南向城市主干道麓山大道开设主门，学员宾馆向西侧城市道路开门，避免对教学楼的干扰，同时通过中部的文体活动区和就餐区作为联系纽带，与教学楼相连接，实现住宿、就餐、教学的一体化设置。

同时用地东侧结合现状绿丘的自然环境，布置景观绿地、运动场和绿茵停车场，环境优雅，减少了对建筑的干扰。北侧用地保留现状鱼塘，进行改造，形成可供学员休憩散步以及相互交流的，环境优美的后湖景观区。

Side Elevation 轴立面图

Side Elevation 轴立面图

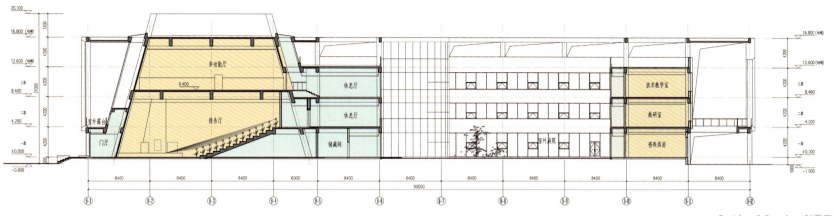

Sectional Drawing 剖面图

Sectional Drawing 剖面图

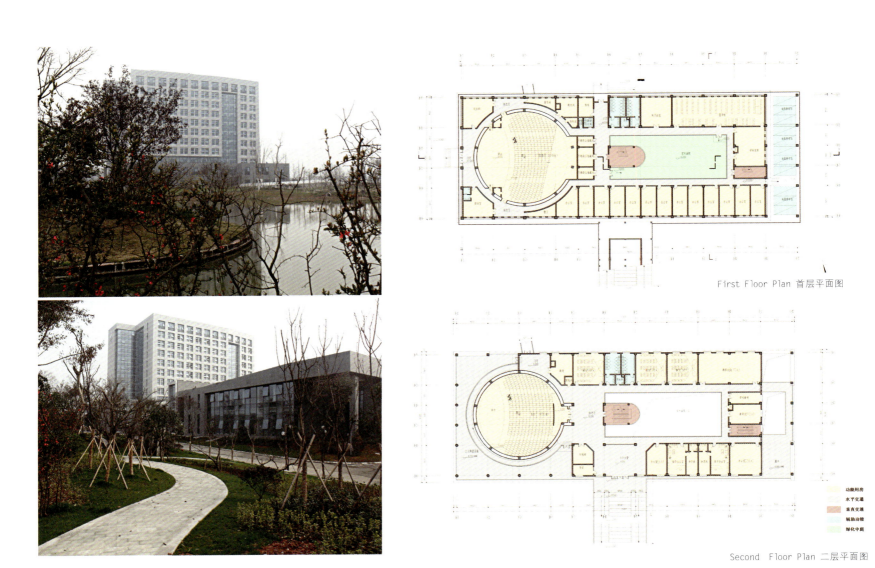

First Floor Plan 首层平面图

Second Floor Plan 二层平面图

160

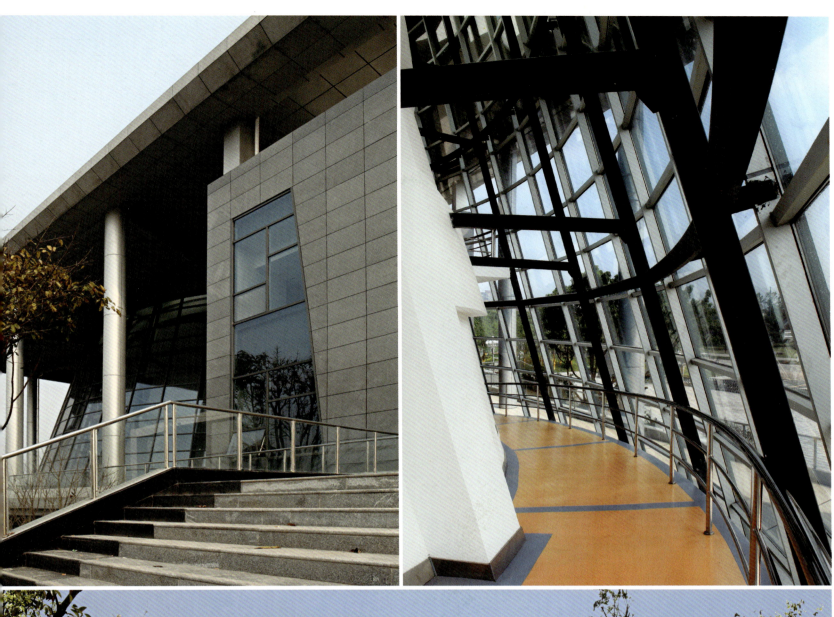

KEY WORDS 关键词

Square Mass 方形体量
Dignified and Stable 端庄稳重
Cold Gray Stone 冷灰色石材

Liaoning College of Communications Library
辽宁交通高等专科学校图书馆

FEATURES 项目亮点

The all cold gray stone exterior design makes the building seamless and natural and at the same time unifies with the gym and main teaching building in external color.

外饰面设计全部为冷灰色石材，使建筑浑然一体，宛自天成，又与已建成的体育馆及主教学楼在外部空间色彩上统一协调。

Location: Shenyang City, Liaoning Province, China
Architectural Design: Architectural Design and Research Institute of Tianjin University
Total Land Area: 12,000 m²
Building Basement Area: 4,330 m²
Total Building Area: 19,270 m²
Green Area: 4,200 m²
Plot Ratio: 1.6
Greening Rate: 35%

项目地点：中国辽宁省沈阳市
设计单位：天津大学建筑设计研究院
总用地面积：12 000 m²
建筑基底面积：4 330 m²
总建筑面积：19 270 m²
绿地面积：4 200 m²
容积率：1.6
绿化率：35%

Architectural Design

On the building façade, the design takes the dignified and elegant outlook as priority, emphasizing the overall sculptural sense. With a full square shape, the building sketches out the overall outline using subtraction, makes a large contrast between the void and solid, the lateral and vertical, and according to functional needs makes some subtle changes on each façade. Therefore in the framework the building is unified as a harmonious whole. The all cold gray stone exterior design makes the building seamless and natural and at the same time unifies with the gym and main teaching building in external color.

建筑设计

在建筑立面造型上以端庄稳重、典雅大方为前提，强调整体的雕塑感。建筑在一个完整的方形体量中，用减法勾勒出整体轮廓，采用大的虚实对比、横竖对比，但每个立面根据功能需要，都有一些细微的变化，在大的构架中统一成为一个和谐的整体。外饰面设计全部为冷灰色石材，使建筑浑然一体，宛自天成，又与已建成的体育馆及主教学楼在外部空间色彩上统一协调。

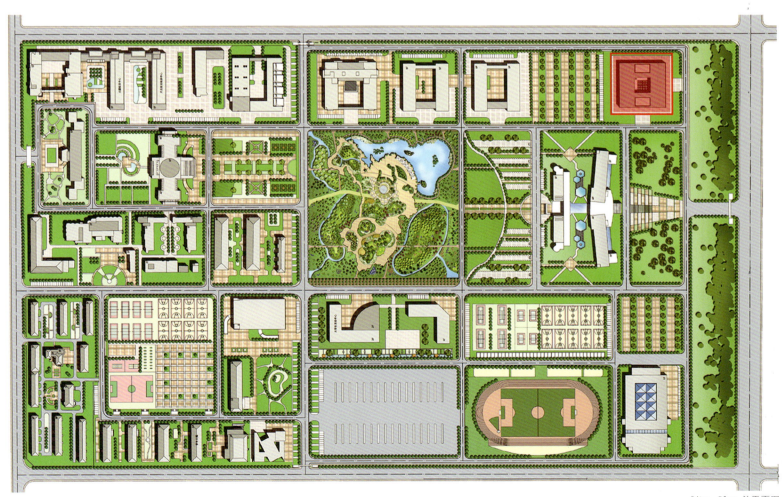

Site Plan 总平面图

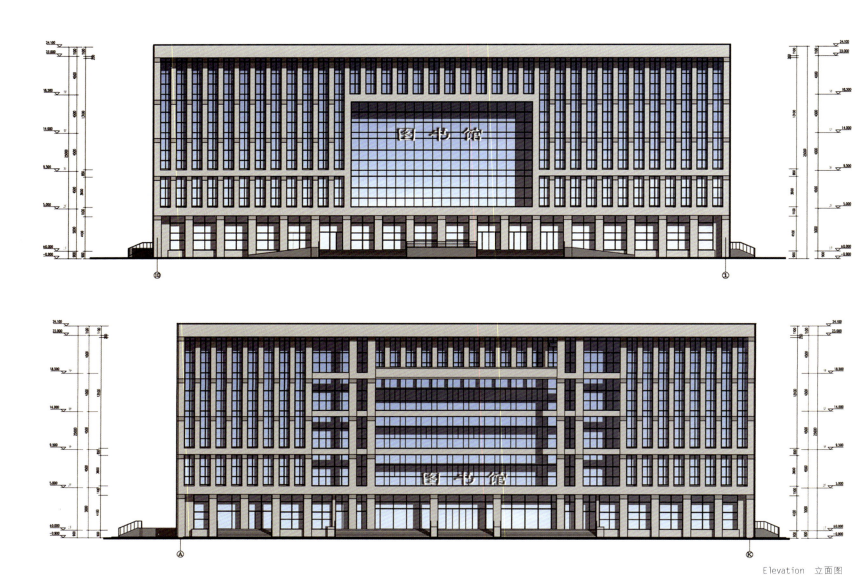

Elevation 立面图

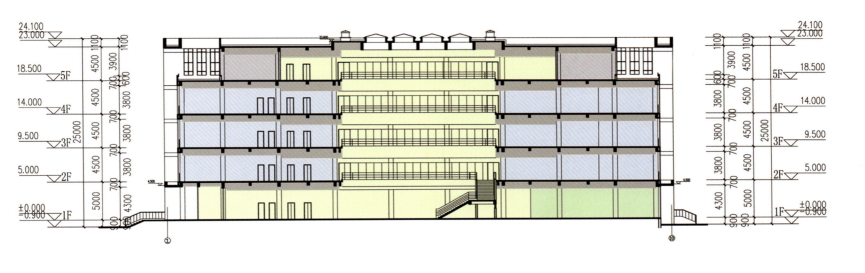

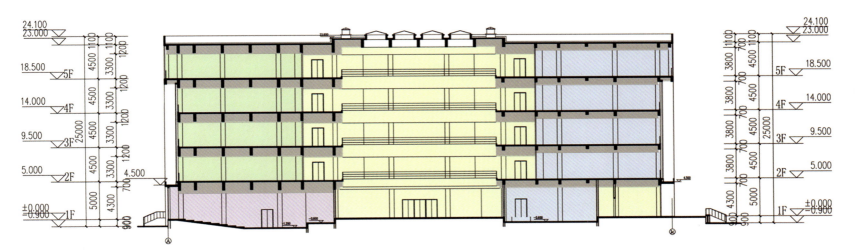

Sectional Drawing 剖面图

First Floor Plan 首层平面图

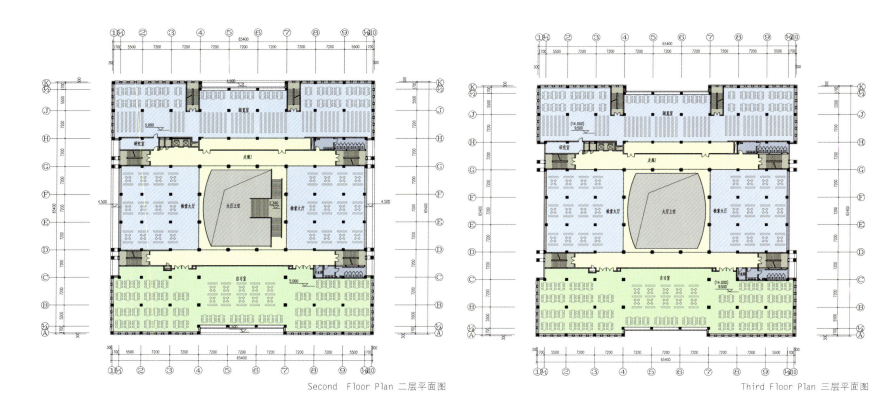

Second Floor Plan 二层平面图

Third Floor Plan 三层平面图

KEY WORDS 关键词

Green and Ecological 绿色生态
Glass Wall 玻璃墙面
Slope Roof 斜坡屋面

Library of Shandong Union Management Cadres College
山东省工会管理干部学院图书馆

FEATURES 项目亮点

The roofs of 2nd -5th floors of the building with setbacks, develop into a holistic slope roof, which leads sunshine to northern side of the building, and makes the dark side of the building bright; the green and ecological slope roof make the building full of vitality.

建筑二层至五层的屋面层层退台，形成一个整体性的斜坡屋面，把阳光引入建筑的北面，使建筑的暗面也亮起来，绿色生态的斜屋面也让建筑充满了活力。

Location: Jinan, Shandong, China
Architectural Design: Shandong Jianda Architectural Planning and Design Institute
Floor Area: 18,768 m²
Building Height: 24 m
Completion：2011

项目地点：山东省济南市
设计单位：山东建大建筑规划设计研究院
建筑面积：18 768 m²
建筑高度：24 m
竣工时间：2011 年

Architectural Design

The northern side of the library directly faces the main entrance of the campus. The roofs of 2nd-5th floors with setbacks, develop into a holistic slope roof, which leads sunshine to northern side of the building, and makes the dark side of the building become bright, the green and ecological slope roof make the building full of vitality. Two-storey angled-out glass wall runs from east to west, adds some thoughtfulness and calmness. A large area of water in the northern side as well as the library create a quiet and elegant atmosphere.

建筑设计

图书馆北面正对校区主入口，将二层至五层的屋面层层退台，形成一个整体性的斜坡屋面，把阳光引入建筑的北面，使建筑的暗面也亮起来，而绿色生态的斜屋面也让建筑充满了活力。两层高的外斜玻璃墙面贯穿东西，为图书馆增添了几分凝重与沉稳。北面大面积的水面，也与图书馆共同营造出安静优雅的环境氛围。

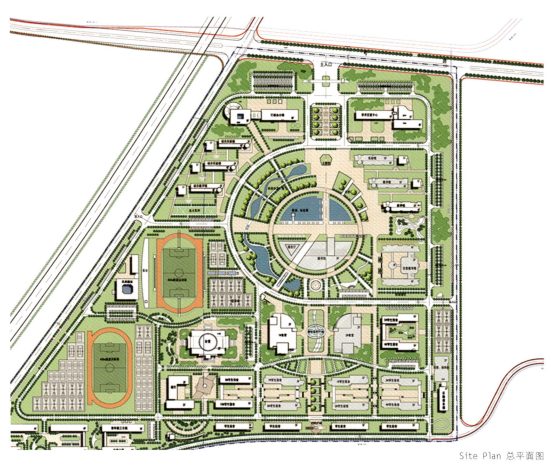

Site Plan 总平面图

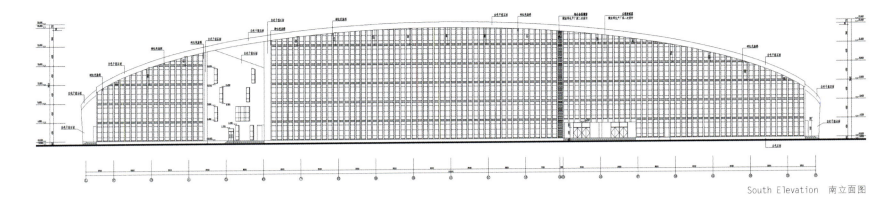

South Elevation 南立面图

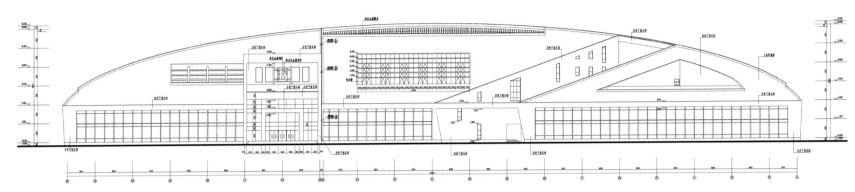

North Elevation 北立面图

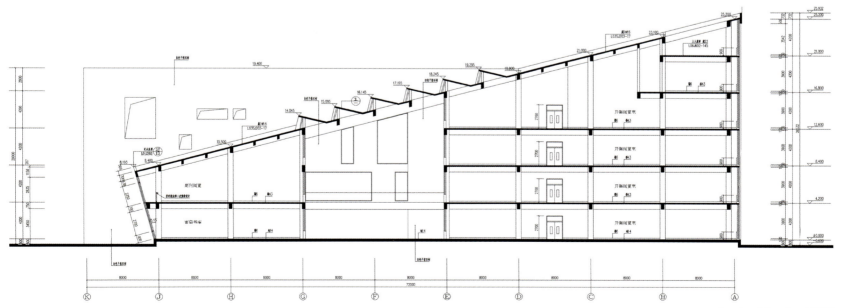

Sectional Drawing 剖面图

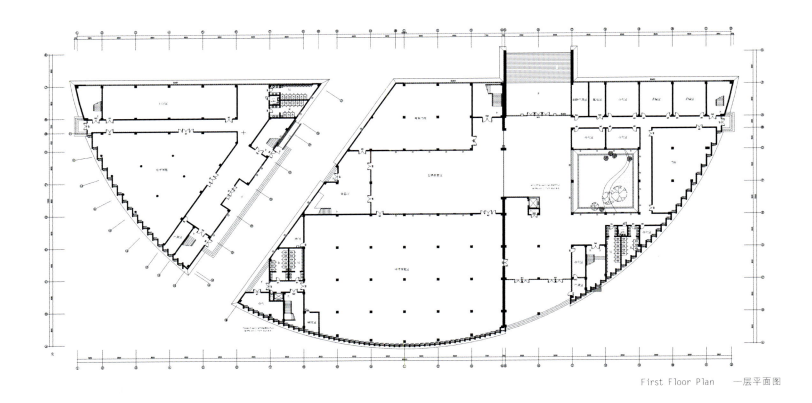

First Floor Plan　一层平面图

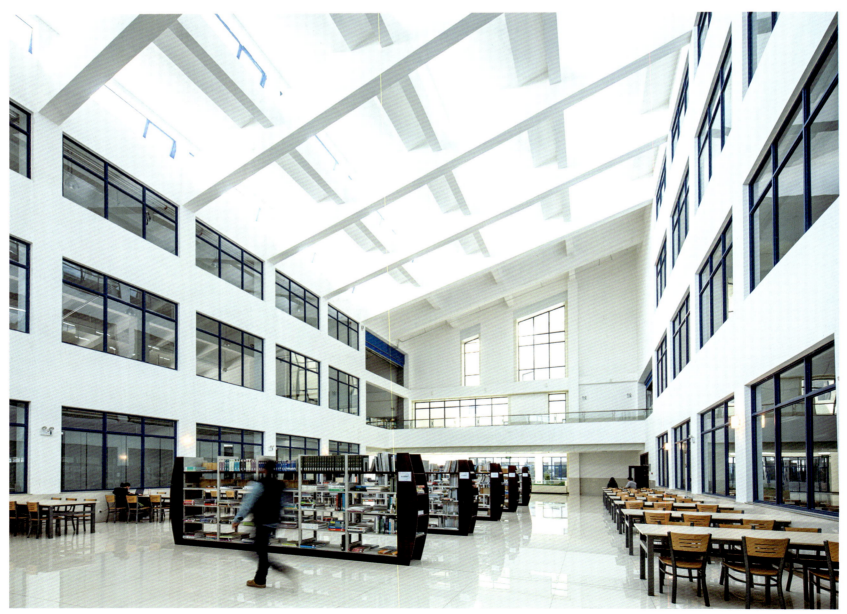

KEY WORDS
关键词

Dry-Sticking Granite 花岗岩干挂
Ceramic Decorative Finishing 陶砖饰面
Glass Curtain Wall 玻璃幕墙

Library of Changchun University of Chinese Medicine
长春中医药大学图书馆

FEATURES
项目亮点

The square large-scale colonnade of the building creates solemn academic image, the angled walls in eastern and western sides look like unfolded book pages, both of which highlight the strong atmosphere of academic palace.

建筑以方正的大尺度柱廊来营造庄严的学术圣殿形象，东西侧斜墙仿若展开的书页，一同烘托出学术殿堂的浓厚氛围。

Location: Changchun, Jilin, China
Architectural Design: Tsinghua University Architectural Design and Research Institute Co., Ltd
Site Area: 28,800 m²
Total Floor Area: 30,717 m²
Completion: 2010

项目地点：中国吉林省长春市
设计单位：清华大学建筑设计研究院有限公司
占地面积：28 800 m²
总建筑面积：30 717 m²
竣工时间：2010 年

Architectural Design

The square large-scale colonnade of the building creates solemn academic image, it also indicates the concept of "Si Ku" in the overall planning of the campus. The steps at the entrance look like "the road to knowledge", the gate of the main entrance symbolizes "the gate open to knowledge", the angled walls in eastern and western sides look like unfolded book pages, both of which highlight the strong atmosphere of academic palace. Regional characteristics and energy saving of the building are fully considered, and the building features square shape and compact layout. Outside enclosing structure mainly employs brick wall, combining with the glass curtain wall that partly faces south, for the sake of adapting to cold climate in the north China. The façade design of east-west facing reading room presents southern inclination of 45 degrees, which can avoid sun shine, and at the same time shake off the sense of monotony of flat wall; it looks like unfolded pages, which indicates the properties of the building. Three interior courtyards of different sizes meet the requirements of day lighting and ventilation, meanwhile, they enrich the space level of the building. The main courtyard facing south with layers of setbacks, tries to achieve best day lighting for the rooms in northern side.

建筑设计

建筑以方正的大尺度柱廊来营造庄严的学术圣殿形象，也暗合了校园总体规划中"四库"的概念。入口处大台阶仿佛"书山之路"，主入口门廊象征"知识圣殿的大门"，东西侧斜墙仿若展开的书页，一同烘托出学术殿堂的浓厚氛围。充分考虑地域特色和建筑节能要求，体形方正，布局紧凑。外围护结构以实墙面为主，结合局部南向玻璃幕墙，以适应北方严寒气候。东西向阅览室外墙设计成45度南向倾斜，有效避免东西晒，同时打破平直墙面的单调感，仿佛展开的书页，暗合了建筑物的使用性质。三个不同尺度的内庭院设计，满足了大进深建筑采光通风要求，同时丰富了建筑空间层次。主庭院南向层层退台，为庭院北侧用房争取最大采光。

Site Plan 总平面图

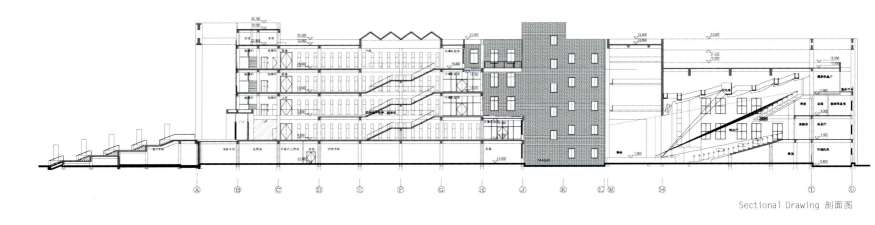
Sectional Drawing 剖面图

East Elevation 东立面图

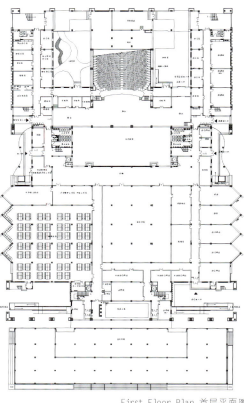

First Floor Plan 首层平面图

KEY WORDS 关键词

Open Space 开放空间
Lighting Atrium 采光中庭
Detail 细部

North Campus of Zhoukou Normal University Library Management Center
周口师范学院北校区图书馆管理中心

FEATURES 项目亮点

Library and administrative office building are relatively independent, but connected each other through corridor, which not only meets the walking demands from east to west for teachers and students, but also provides a individual academic space.

行政办公楼与图书馆建筑既相对独立，又通过连廊相互联系，不仅可以满足东西两区师生的步行来往需求，更为其提供了一个别具氛围的学术交流空间。

Location: Zhoukou, Henan, China
Architectural Design: Design and Research Institute of Zhengzhou University
Total Floor Area: 33,140 m²
Completion: 2011

Overview

This project is located in North Campus of Zhoukou Normal University, including library and administration office building, with 2 million copies of collections. The library building with 5 floors is as high as 21.13 m, and the administration office building with 6 floors is as high as 23.93 m.

项目地点：中国河南省周口市
设计单位：郑州大学综合设计研究院
总建筑面积：33 140 m²
竣工时间：2011年

项目概况

本工程位于周口师范学院北校区，其主要为图书馆和行政办公楼，藏书量共200万册。其中图书馆建筑共五层，建筑高度为21.13 m；行政办公楼六层，建筑高度为23.93 m。

Planning and Layout

The site of this project is located in central position of north-south axis. The teaching buildings are mainly multi-storey ones, which spread following the existing topography, and develop into continuous and rhythmical architectural landscape. As the project is carried out after the overall planning campus completed, this newly built library management center coordinate with the surrounding environment in the layout and the overall planning, and it develops into "Triangle" shape through combining teaching buildings on both sides of the campus and lab buildings, forming a space environment in campus square.

规划布局

该管理中心基地位于学校南北主轴线的中心位置。南北轴线两侧的教学楼设计均以多层为主，顺应原有地形铺展，形成富有韵律、连续的建筑景观。由于本项目是在校园整体规划实施基本完成的基础上进行的，新建的图书馆管理中心在布局上与整体规划，与周边环境相协调，与校园前区两侧的教学、实验楼群呈"品"字形，形成学校前区广场的空间环境。

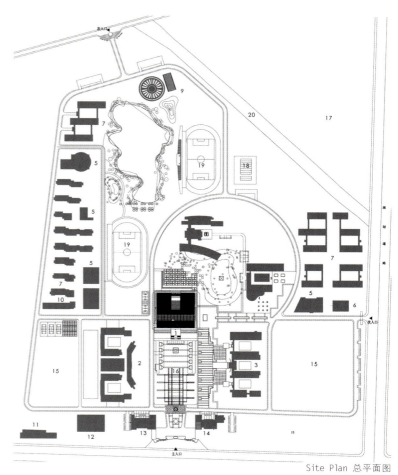

Site Plan 总平面图

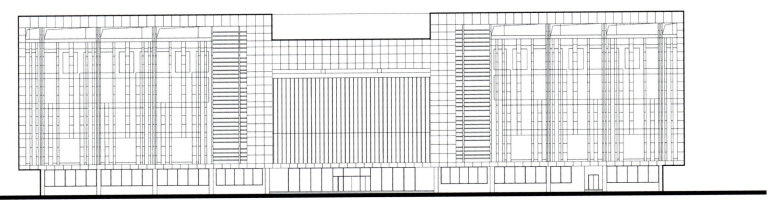

North Elevation 北立面图

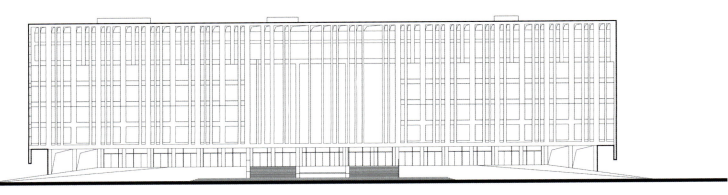

South Elevation 南立面图

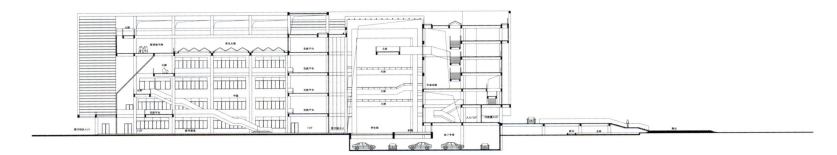

Sectional Drawing 剖面图

Functional Division

The project consists of library and administrative office building. As the site is located in the central position of the campus overall planning, a transversal road across the campus from east to west exists in the site. In the design, the administration office building and the library are divided into southern and northern parts, there is a linear open space between which. These two buildings are relatively independent, but connected each other through corridor, which not only meets the walking demands from east to west for teachers and students, but also provides a individual academic space.

功能分区

本项目包括图书馆和行政办公楼两大主要功能。由于基地位于学校整体规划的中心位置，基地现状有一条贯穿东西的校园横向主要道路。在设计中，行政办公楼与图书馆的南北分区，在其中布置了一条线形开放空间。这样两组建筑既相对独立，又通过连廊相互联系。不仅可以满足东西两区师生的步行来往需求，更为其提供了一个别具氛围的学术交流空间。

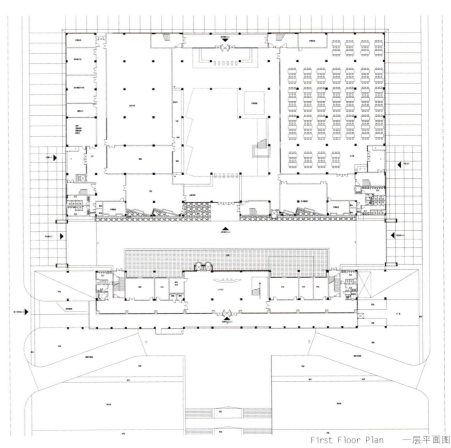

First Floor Plan 一层平面图

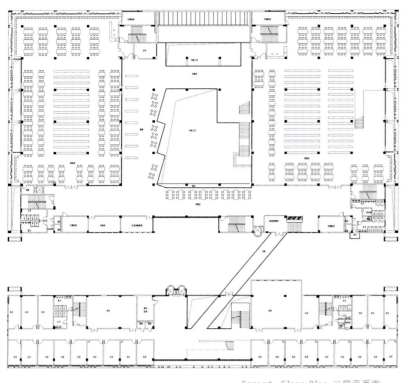
Second Floor Plan 二层平面图

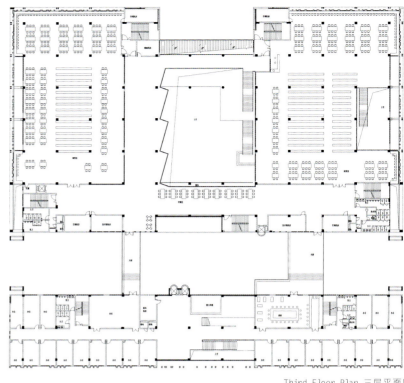
Third Floor Plan 三层平面图

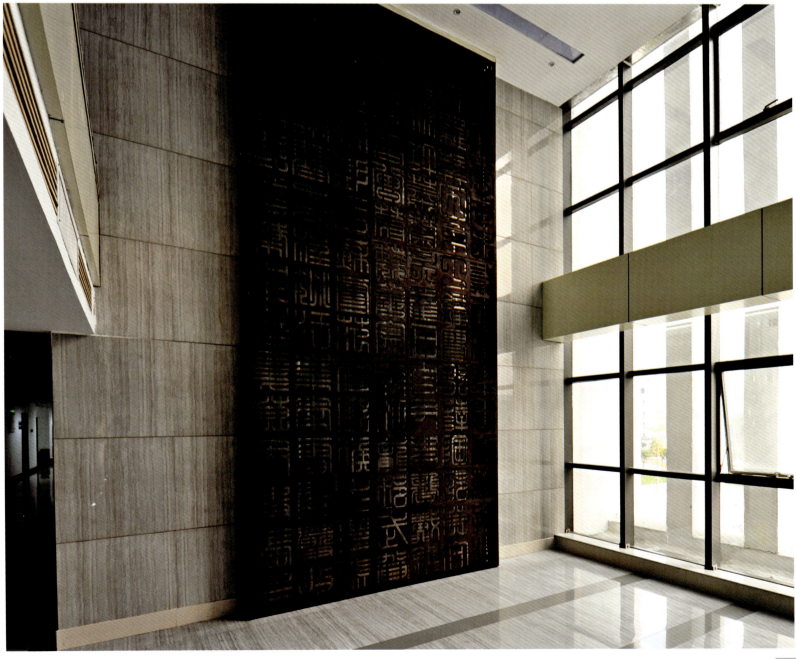

Transportation Building
交通建筑

- Reasonable Organization
  布局合理

- Modern Technology
  现代技术

- Great Convenience
  便利生活

- Green Building
  绿色建筑

KEY WORDS 关键词

Curved Form 曲线形态
Dendritic Pattern 树枝状造型
White Shape 白色形体

Daqing Highway Passenger Terminal
大庆市公路客运枢纽站

FEATURES 项目亮点

Revolved around "ice and snow culture", designers draw on the characteristics of Northeast China snow landscape. Rolling and interlaced white shape looks like a winter hill standing on the earth, creating a graceful curved form.

建筑形体塑造以"冰雪文化"为入手点，借鉴了东北地区雪原地貌的特点，起伏交错的白色形体犹如冬季的山丘一样矗立在大地之上，创造了优美的曲线形态。

Location: Daqing, Heilongjiang, China
Architectural Design: Tianchen Architectural Design Co., Ltd.
Total Land Area: 67,579 m²
Main Building Area: 26,729 m²
Plot Ratio: 0.37

项目地点：中国黑龙江省大庆市
设计单位：哈尔滨天宸建筑设计有限公司
总用地面积：67 579 m²
主体建筑面积：26 729 m²
容积率：0.37

Overview

The 57.4 m main building is comprised of a 2-storey passenger station building and a 14-storey information center. The passenger station highlights openness and permeability in its interior space, and the shared hall runs through the three floors and links up each functional space, providing a favorable flow line and sight for the passengers.

项目概况

项目主体建筑由3层的客运站站房和14层的信息中心楼组成，建筑高度57.4 m。客运站内部空间突出了开敞性和通透性，设置了贯通三层的共享大厅，将各部分功能空间联系在一起，为旅客提供了良好的流线与视线。

Architectural Design

Revolved around "ice and snow culture", designers draw on the characteristics of Northeast China snow landscape. Rolling and interlaced white shape looks like a winter hill standing on the earth, creating a graceful curved form. In terms of façade detail, steel structural elements are organized in a dendritic pattern, which enriches the depth of the façade and strengthens the design concept as well. The building skin is comprised of white aluminium sheet and glass curtain wall, and the sheet interweaves with glass flexibly so as to reach a perfect and organic combination.

建筑设计

建筑形体塑造以"冰雪文化"为入手点，借鉴了东北地区雪原地貌的特点，起伏交错的白色形体犹如冬季的山丘一样矗立在大地之上，创造了优美的曲线形态。建筑的立面细节通过钢结构构件形成了树枝状的造型，丰富了立面层次，也强化了设计理念。建筑的表皮由白色铝板和玻璃幕墙组成，铝板与玻璃灵活穿插，通过自由曲线有机地组合在一起。

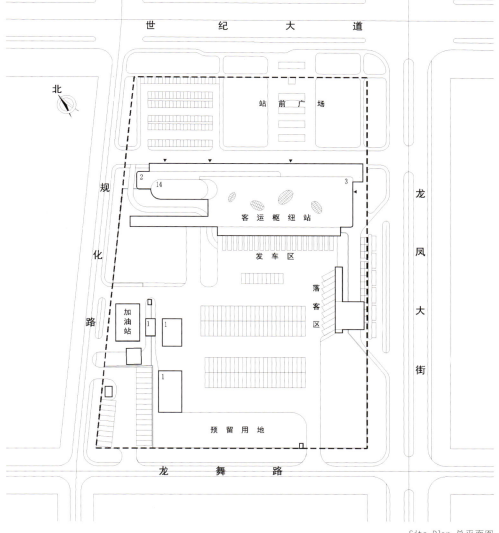

Site Plan 总平面图

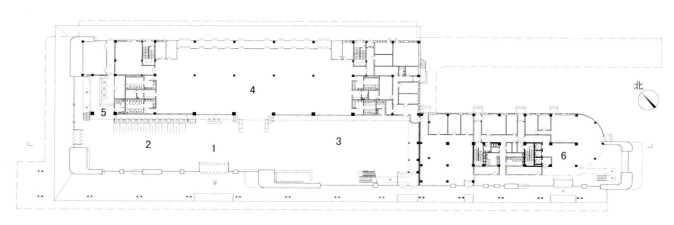

1 客运站门厅　2 售票厅　3 餐厅
4 候车厅　5 行包托运　6 信息中心门厅

First Floor Plan　一层平面图

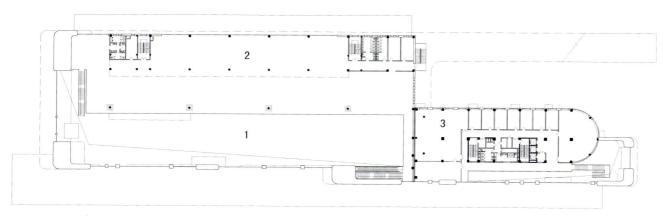

1 共享大厅　2 候车厅　3 信息中心办公

Second Floor Plan　二层平面图

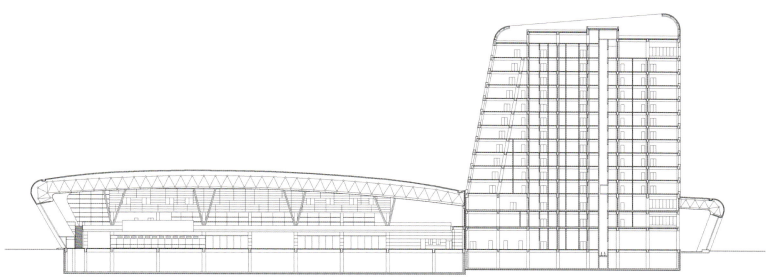

Sectional Drawing 剖面图

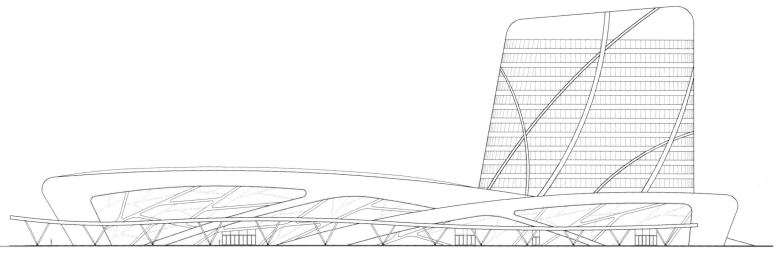

Front Elevation 正立面图

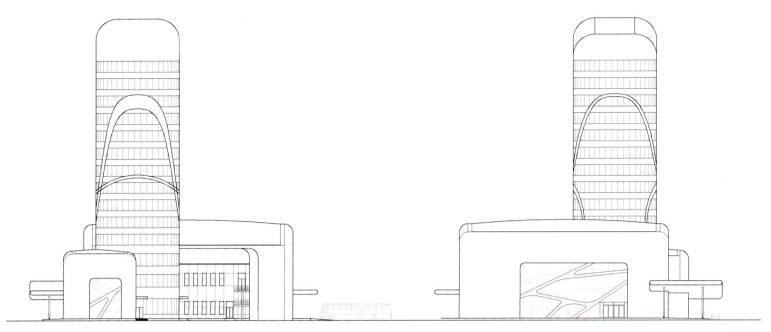

End Elevation 侧立面图

KEY WORDS 关键词

Garden-like Terminal 花园式航站楼
Graceful Shape 造型优美
Roof System 屋面系统

Terminal of Jieyang Chaoshan International Airport
揭阳潮汕机场航站楼及配套工程

FEATURES 项目亮点

In consideration of the passengers' experience and the client's demands, the building is designed in garden style with typical Chaoshan elements and non-linear metallic roof system.

建筑是集当地传统元素、旅客出行体验、业主使用需求于一身的原创建筑，具有精心构筑的花园式航站楼、造型设计融合潮汕地域特色和时代特征、非线性金属屋面系统设计。

Location: Jieyang, Guangdong, China
Architectural Design: Architectural Design and Research Institute of Guangdong Province

项目地点：中国广东省揭阳市
设计单位：广东省建筑设计研究院

Overview

This three-storey terminal is high to 30.17 m, featuring a total floor area of 56,811 m², the building density of 6% and the plot ratio of 0.15. Part of the CIP has four floors and the base covers an area of 22,832 m². The equipment center is 5.5m high. With one floor underground and one above ground, it has a total floor area of 1,941 m².

项目概况

航站楼总建筑面积56 811 m²，建筑密度为6%，容积率为0.15，建筑总高度30.17 m，建筑地上三层（CIP局部四层），建筑基底面积22 832 m²。设备中心总建筑面积1 941 m²，建筑高度5.5 m，建筑地上一层、地下一层。

Architectural Design

In consideration of the passengers' experience and the client's demands, the building is designed in garden style with typical Chaoshan elements and non-linear metallic roof system.

建筑设计

建筑为集当地传统元素、旅客出行体验、业主使用需求于一身的原创建筑，具有精心构筑的花园式航站楼、造型设计融合潮汕地域特色和时代特征、非线性金属屋面系统设计。

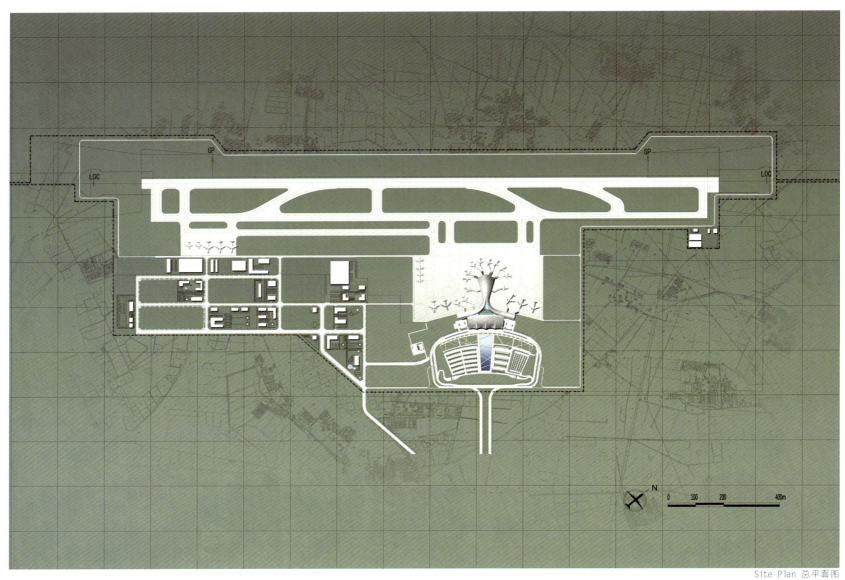

Site Plan 总平面图

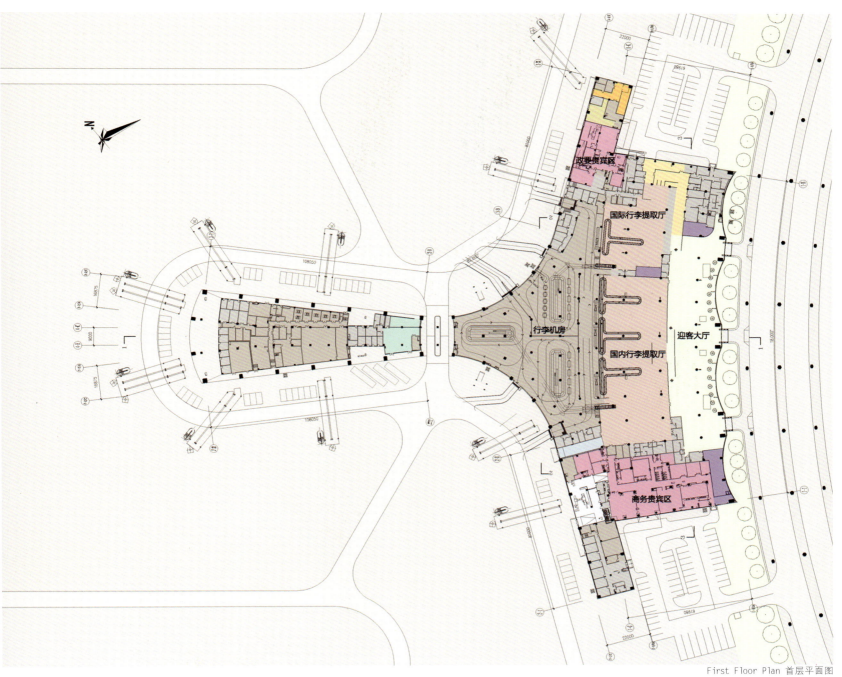

First Floor Plan 首层平面图

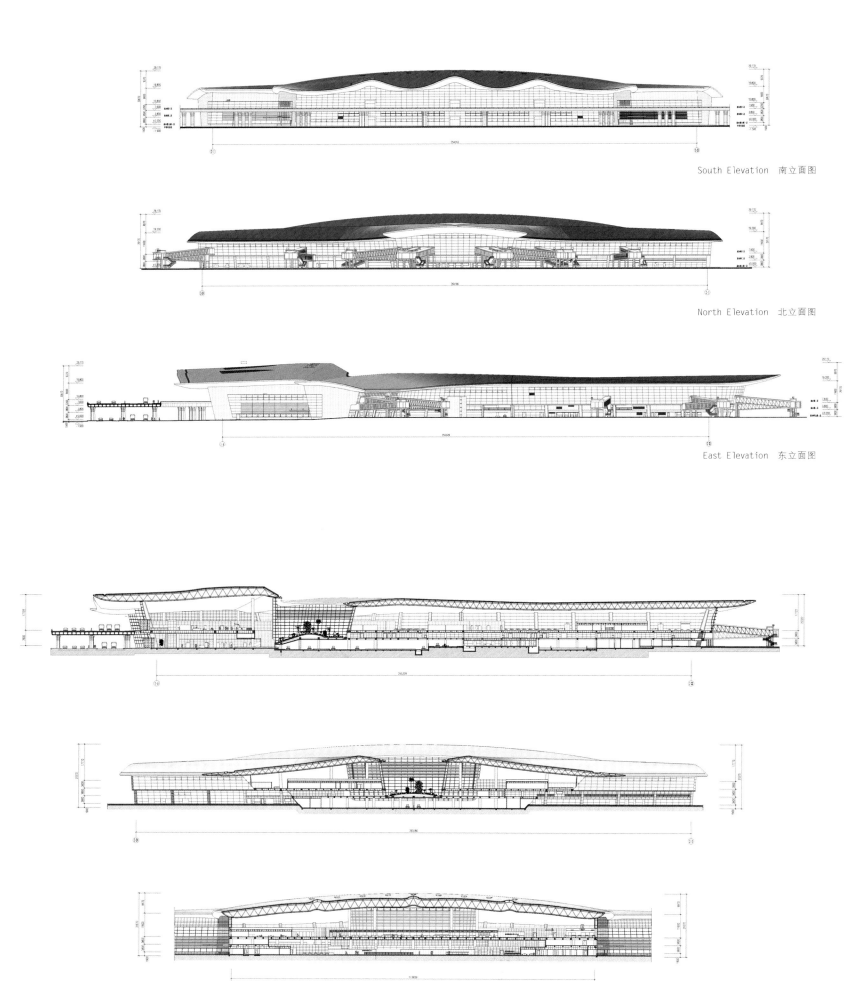

South Elevation 南立面图

North Elevation 北立面图

East Elevation 东立面图

Sectional Drawing 剖面图

KEY WORDS 关键词

Roof System 屋盖体系
Green Courtyard 绿色庭院
Building Materials 建筑材料

Suzhou Automobile North Station Reconstruction Project
苏州汽车北站改建工程

FEATURES 项目亮点

The project adopts modern materials and design methods to express traditional symbols and details, creating a new carrier adapting to the time for traditional architectural culture.

项目用现代的材料和设计手法表达传统符号与细部，为传统的建筑文化塑造一个适应时代的新的载体。

Location: Suzhou, Jiangsu, China
Architectural Design unit: Suzhou Institute of Architectural Design Co., Ltd.
Total Land Area: 32,019 m^2
Total Floor Area: 24,778 m^2
Building Height: 22.8 m
Building Density: 18.38%
Plot Ratio: 0.3749
Greening Ratio: 16.5%
Completion: 2011

项目地点：中国江苏省苏州市
设计单位：苏州设计研究院股份有限公司
总用地面积：32 019 m^2
总建筑面积：24 778 m^2
建筑高度：22.8 m
建筑密度：18.38%
容积率：0.3749
绿化率：16.5%
竣工时间：2011年

Overview

Suzhou Automobile North Station is located at the intersection of Xihui Road and Suyu Road in northern Suzhou's ancient city. It has three layers above the ground and one layer underground, and a total building height of 22.8 m.

项目概况

苏州汽车北站位于苏州古城区北部，西汇路和苏虞路交叉口。建筑地上三层，地下一层，建筑高度22.8 m。

Planning and Layout

This half C shaped building lies in the southwest of the project base, sets aside the city square along the Suyu Road and Xihui Road, and encloses an internal parking lot. The main station building spreads along the Xihui Road like a straight line; crowds and traffic from the ground station are evacuated from Xihui Road. Near Suyu Road and Xihui Road, there is a major pedestrian entrance and sunken square. Cars and buses enter from the east to the east, and depart from the ramp in the Beihuan Road after picking up all the passengers.

规划布局

半C字形建筑设于基地西南，沿苏虞路和西汇路留出城市广场，围合出内部的客车停车场。建筑主体站房"一"字形沿西汇路展开，地面到站人流和车流均由西汇路疏散。临苏虞路、西汇路口设主要人行入口和下沉式广场。场内客运车辆及公交车由东面西汇路进，最后上客后由北环路上专设的匝道驶出。

Site Plan 总平面图

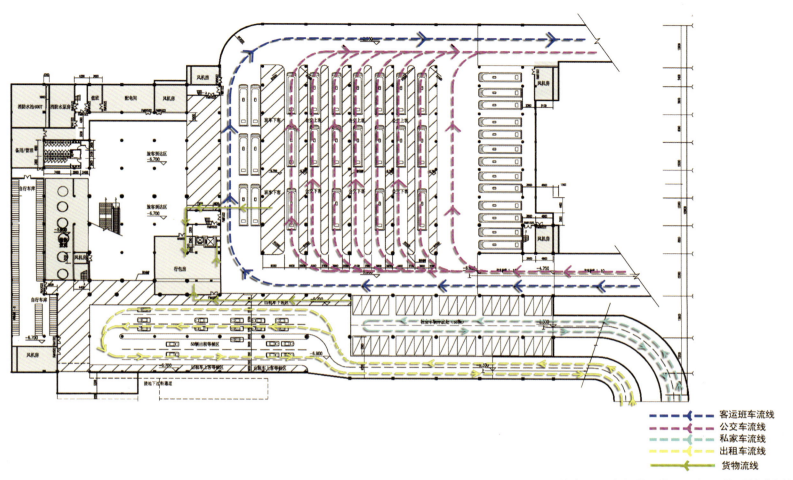

Underground Traffic Flow Analysis 地下层车流分析图

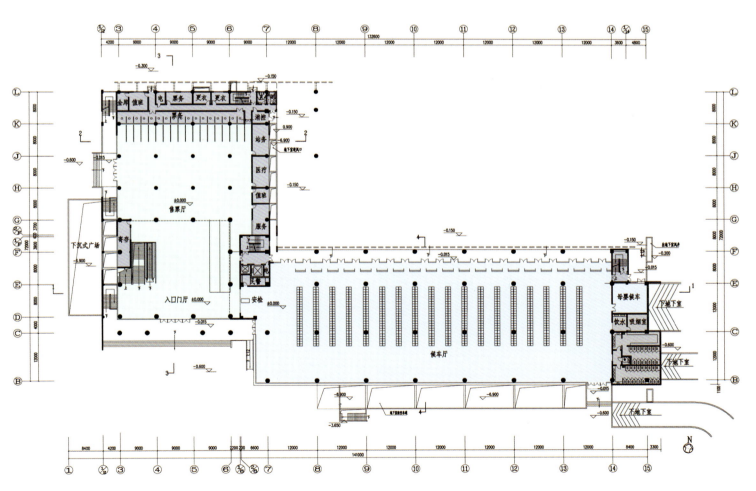

First Floor Plan 一层平面图

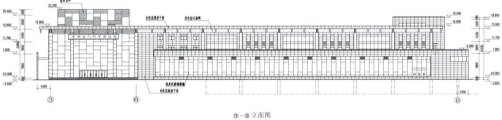

Elevation 立面图

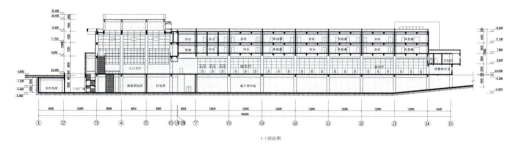

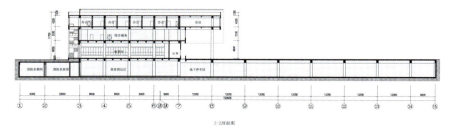

Sectional Drawing 剖面图

Architectural Design

The building is enclosed to be courtyard space with roof and walls, forming an inward spatial pattern; inside the architectural space, there are multiple greening gardens, leading in natural elements such as the sun, wind, water and greening to create a bright space with rich levels. The roof system has clear structure with far-reaching eave, which is exposed like grey rafter secondary beam; pane windows and ornamental perforated window keep a moderate proportion, and the white walls apply delicate black steel channel to make coping by using modern materials and design methods to express the traditional symbols and details, creating a new carrier adapting to the time for traditional architectural culture.

建筑设计

建筑以顶盖、墙垣等围合成院落空间，营造成内向性的空间形态，在建筑空间内部穿插了多个绿化庭院，将阳光、风、水、绿化等自然元素引入，并形成层次丰富的明暗空间。屋盖体系结构层次清晰，屋檐出挑深远，外露如椽子般韵律的灰色次梁；方格窗、漏窗比例适中，白墙上冠以精致的黛色槽钢压顶，用现代的材料和设计手法表达传统符号与细部，为传统的建筑文化塑造一个适应时代的新的载体。

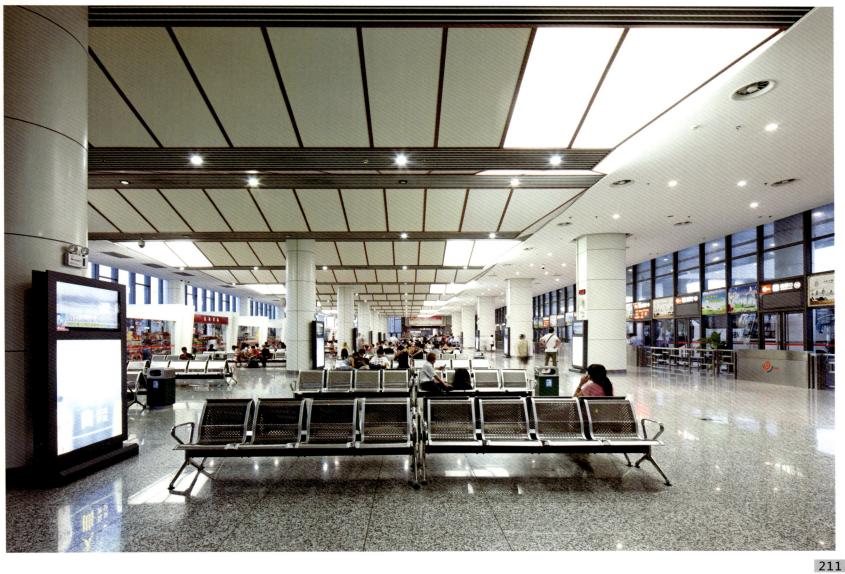

KEY WORDS 关键词	**Elegant Molding** 造型简洁
	Reasonable Arrangement 布局合理
	Colorful Facade 立面丰富

Suzhou Rail Transit Line 1 Control Center

苏州市轨道交通一号线工程控制中心大楼

FEATURES 项目亮点

The whole building is elegantly designed to keep harmonious with the surroundings, and the podium well echoes the main tower.

建筑整体布局与周边环境和谐统一，造型简洁明朗，主楼与裙楼的两部分相互呼应。

Location: Suzhou, Jiangsu, China
Architectural Design: Suzhou Institute of Architectural Design Co.,Ltd.
Total Land Area: 8,211 m^2
Total Floor Area: 57,697 m^2
Building Height: 99.9 m
Building Density: 34.6%
Plot Ratio: 5.30
Greening Ratio: 13%
Completion: 2011

项目地点：中国江苏省苏州市
设计单位：苏州设计研究院股份有限公司
总用地面积：8 211 m^2
总建筑面积：57 697 m^2
建筑高度：99.9 m
建筑密度：34.6%
容积率：5.30
绿化率：13%
竣工时间：2011 年

Overview

Boasting a total floor area of 57,697 m^2, Suzhou Rail Transit (SRT) Line 1 Control Center is located at the intersection of Ganjiang Road and Guangji South Road. Line 1 and Line 2 just meet here. The tower is high to 23 floors and the podium has 5 floors. There are also 3 floors underground. It is the center for the supervision, control and management of the SRT lines.

项目概况

苏州轨道交通控制中心大楼位于干将路与广济南路交叉口，即1号线、2号线交会处，总建筑面积57 697 m^2，其中，地下3层，地上主楼23层、裙楼5层，具有苏州轨道交通线路的集中监控、调度、指挥、管理等功能。

Site Plan 总平面图

Planning

The site is longer from east to west than that from south to north, with Ganjiang Road to the south and Guangji Road on the east. Entrances for motor vehicles are set at the northeast corner on Guangji Road and southwest corner on Ganjiang Road, forming an inner ring road for fire & emergency. There's another entrance for pedestrian on Ganjiang Road. Accordingly, the building is designed in rectangular shape and extends from south to north with a large area plaza in the south. While in the north, there is an arena for firefighting and an entry for staff. Floors 1—7 are the operation control center (OCC), floors 8—23 are used for administrative offices and operation, and the three underground floors are used for parking and equipment rooms. The whole building is designed elegantly to keep harmonious with the surroundings, and the podium well echoes the main tower.

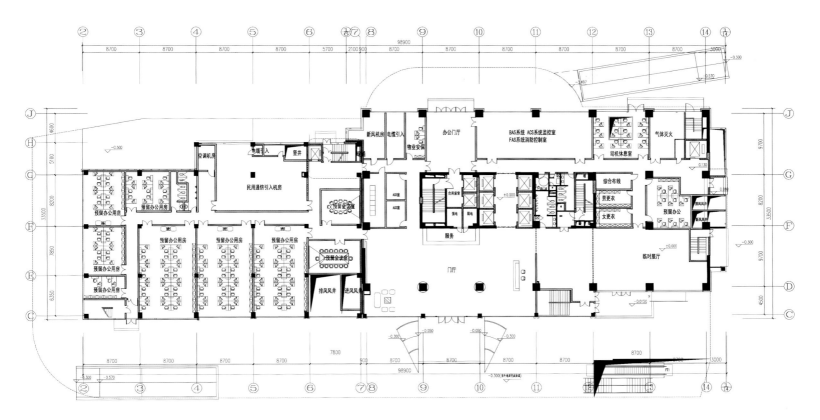

First Floor Plan 一层平面图

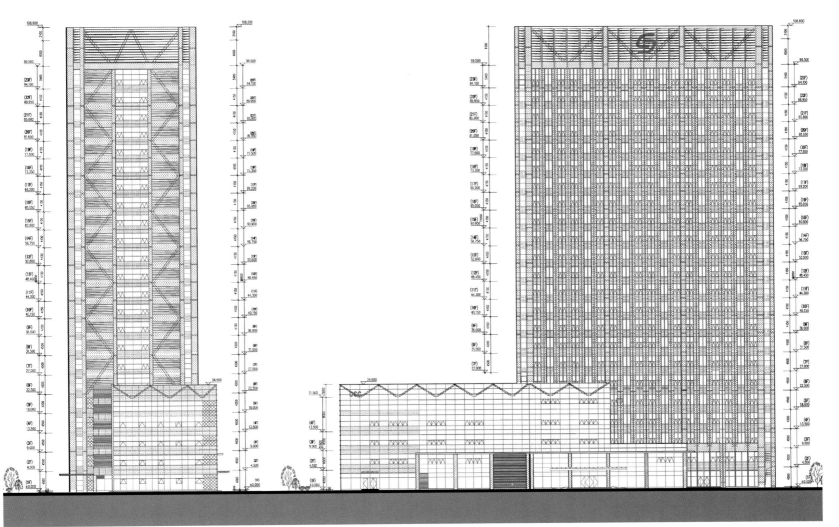

Elevation 立面图

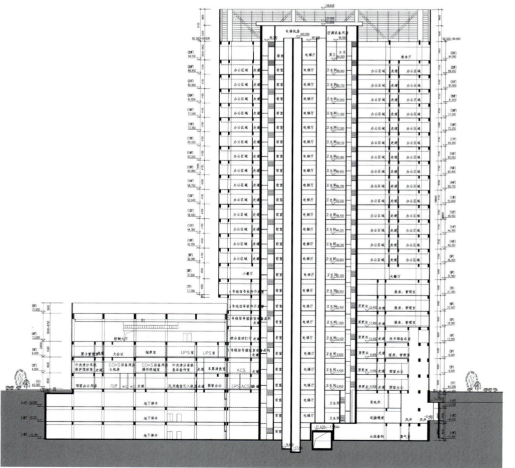

Sectional Drawing 剖面图

项目规划

基地东西方向比较长，南北方向比较短，用地南邻干将路，东接广济路。基地的东北角和西南角分别在广济路和干将路上设一机动车出入口，在基地内形成消防环路。在干将路上另设一人员出入口。根据地形和出入口的设计情况建筑大体呈南北走向，南面根据规划要求和建筑功能设置大面积的集散广场。北面为建筑的消防登高场地和内部人员入口。建筑分为裙楼和主楼两部分，1~7层为轨道交通控制中心OCC用房。8~23层为轨道公司行政办公和运营部分用房。地下3层为大楼停车和辅助设备用房。建筑整体布局与周边环境和谐统一，造型简洁明朗，主楼与裙楼的两部分相互呼应。

KEY WORDS 关键词

Three-dimensional Space 立体空间
Stone Curtain Wall 石材幕墙
Local Design Style 地域设计风格

Turpan Jiaohe Airport Terminal
吐鲁番机场航站楼工程

FEATURES 项目亮点

Designed with one floor, the Terminal features clear and compact space arrangement. The interior and exterior design are simple and elegant in the same style, providing pleasant space experience for the passengers.

航站楼为一层式，功能分区一目了然，内外装修设计繁简适当，风格统一、空间流畅，为乘客提供愉悦的空间体验。

Location: Turpan, Xinjiang Uygur Autonomous Region, China
Architectural Design: Xinjiang Institute of Architectural Design and Research
Land Area: 164,000 m²
Total Floor Area: 5,300 m²
Plot Ratio: 1.18
Completion: 2010

项目地点：中国新疆维吾尔自治区吐鲁番市
设计单位：新疆维吾尔自治区建筑设计研究院
用地面积：164 000 m²
总建筑面积：5 300 m²
容积率：1.18
竣工时间：2010年

Overview

This Terminal is designed with one floor, including the arrival area, departure area, VIP area and baggage claim area. With clear organization and innovative design, luggage conveyor system running from the check-in counters to the baggage room is set under the passenger arrival hall to ensure a compact space arrangement inside the terminal hall. VIP room and the baggage room are set on two ends with independent spaces for future development. The interior and exterior are designed to be simple and elegant in the same style, providing pleasant space experience for the passengers.

项目概况

吐鲁番机场航站楼为一层式，分为到达、出发、贵宾以及行李领取区四大部分。功能分区一目了然，其流程设计有创新点，交运行李从值机柜台输送到行李房，是经由旅客到达厅的下层空间，立体交叉到达行李处理间，这就使航站楼地面流程极为紧凑，行李处理集中。贵宾室、行李房各处一端，有各自独立的空间和发展的余地，内外装修设计繁简适当，风格统一、空间流畅，为乘客提供愉悦的空间体验。

Site Plan 总平面图

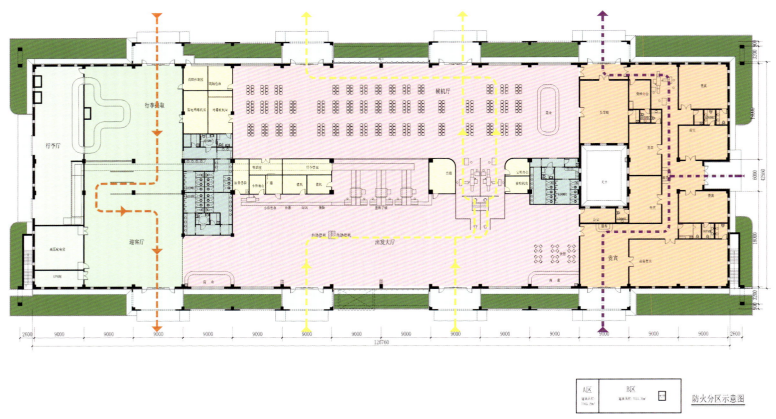

First Floor Plan 一层平面图

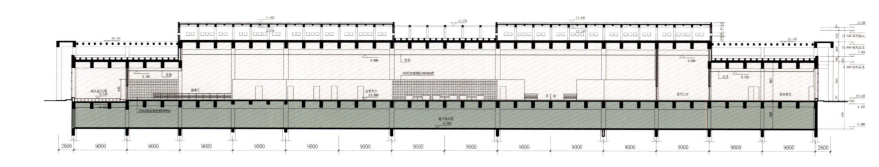

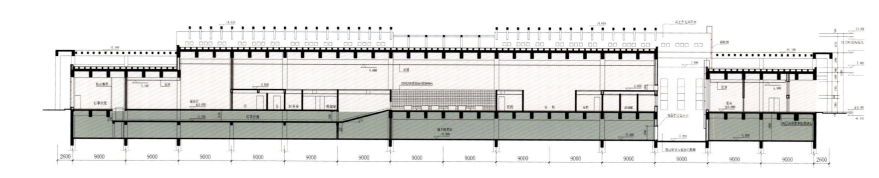

Sectional Drawing 剖面图

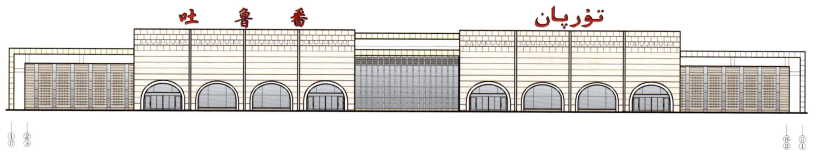

Front Elevation 正立面图

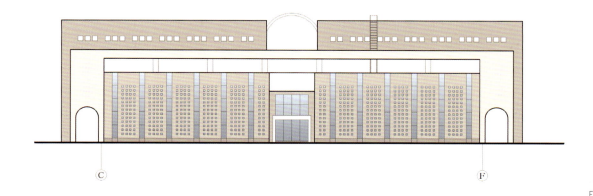

End Elevation 侧立面图

Healthcare Building
医疗建筑

- Elegant Appearance
  外形简约

- Human Scale
  尺度适宜

- People Oriented
  以人为本

- Favorable Environment
  环境良好

KEY WORDS 关键词	Glass Curtain Wall 玻璃幕墙
	Innovative Façade 立面新颖
	Natural Lighting 自然采光

Beijing Chaoyang Hospital Reconstruction & Extension
北京朝阳医院改扩建工程门急诊及病房楼

FEATURES 项目亮点

The project is characterized by the perfect combination of extension and original ward building. The old and new are different in height and connected by ramps and steps. A patio is formed above the top of the existing 5-storey building, which brings natural lighting and ventilation and enriches the spatial order of the building.

扩建病房楼与原有病房楼的巧妙结合，是本工程亮点之一，新老建筑层高不同，用坡道和台阶相连通。利用5层高现状建筑的顶部空间，围合成天井，使建筑体内有自然采光和通风，丰富了建筑的空间秩序。

Location: Chaoyang, Beijing, China
Architectural Design: China IPPR International Engineering Corporation
Land Area: 57,000 m²

项目地点：中国北京市朝阳区
建筑设计：中国中元国际工程公司
用地面积：57 000 m²

Overview

This project is a complex building which accommodates outpatient service, emergency treatment, medical technology and ward, etc. It is 59.8 m and has 13 floors overground and 3 floors underground, and the total floor area is 84,100 m².

项目概况

改扩建工程门急诊及病房楼是一栋集门诊、急诊、医技、病房等为一体的医疗综合楼，总建筑面积84 100 m²，地上13层、地下3层，建筑高度59.8 m。

Architectural Design

In order to improve the problem of land shortage of the old hospital in the urban center, designers emptied bottom two floors of the outpatient building on the east side at the street front entrance, turning the front outdoor square into a buffer area between the building and urban road, which facilitates the circulation of people and vehicle.

The shared hall covered with steel structure glass curtain wall on the west side of the outpatient building is connected with the main entrance hall to create bright and transparent hospital lobby, providing plenty of sunlight and interior & exterior landscape for the main channel and rest space on each floor and beautifying the medical environment. Sections in the outpatient department are relatively independent and connected to a corridor, which allows the patients to find their destinations easily and offers them favorable healthcare space.

The project is characterized by the perfect combination of extension and original ward building. The old and new are different in height and connected by ramps and steps. A patio is formed above the top of the existing 5-storey building, which brings natural lighting and ventilation and enriches the spatial order of the building.

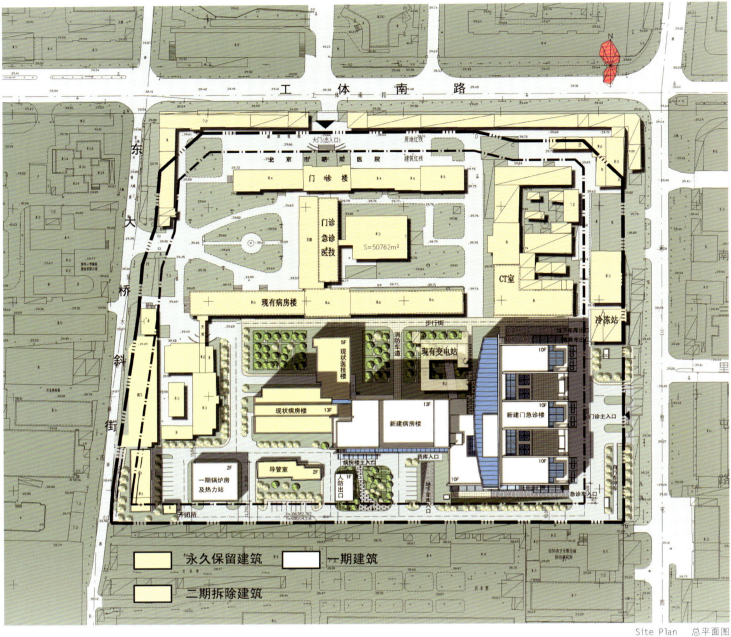

Site Plan 总平面图

建筑设计

为改善城市中心区老医院用地紧迫状况，设计将门诊楼东侧沿街主入口处架空二层，作为医院室外前广场的一部分，使之成为建筑与城市道路之间的缓冲区域，有利于门诊楼前大量人车集散。

门诊楼西侧的钢结构玻璃幕墙共享大厅与主入口大厅贯通，形成通透明亮的医院大堂，给每层门诊主通道及休憩空间以充足的阳光和室内外景观，美化就医环境。门诊部各科室均自成盲端相对独立，并以共用连廊连通，方便患者寻找，为其提供优良医疗空间。

扩建病房楼与原有病房楼的巧妙结合，是本工程亮点之一。新老建筑层高不同，用坡道和台阶相连通。利用5层高现状建筑的顶部空间，围合成天井，使建筑体内有自然采光和通风，丰富了建筑的空间秩序。

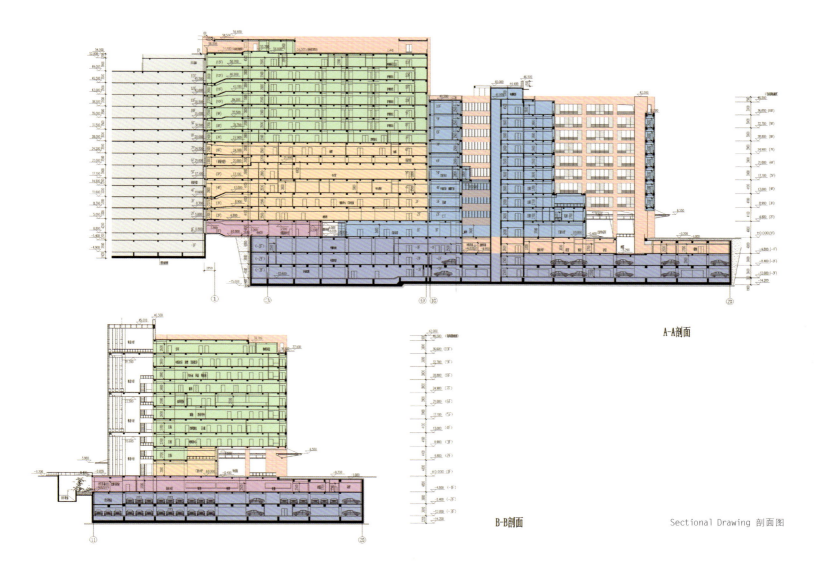

A-A剖面

B-B剖面

Sectional Drawing 剖面图

Façade Design

The façade is innovative, unique, concise and sprightly. The simple and modest dark red bricks and well-structured glass that represent Chaoyang Hospital are used for the exterior wall. The meticulous design allowed the large volume to make the most of the natural lighting and ventilation, thus to avoid great energy consumption in interior space and save operating costs.

立面设计

外立面设计新颖独特、简洁明快。外墙选材朴素，以代表朝阳医院的暗红色外墙砖和规整的玻璃为主。通过精心设计，使大体量建筑尽量利用自然采光、通风，避免建筑内部大量耗能，节省运行费用。

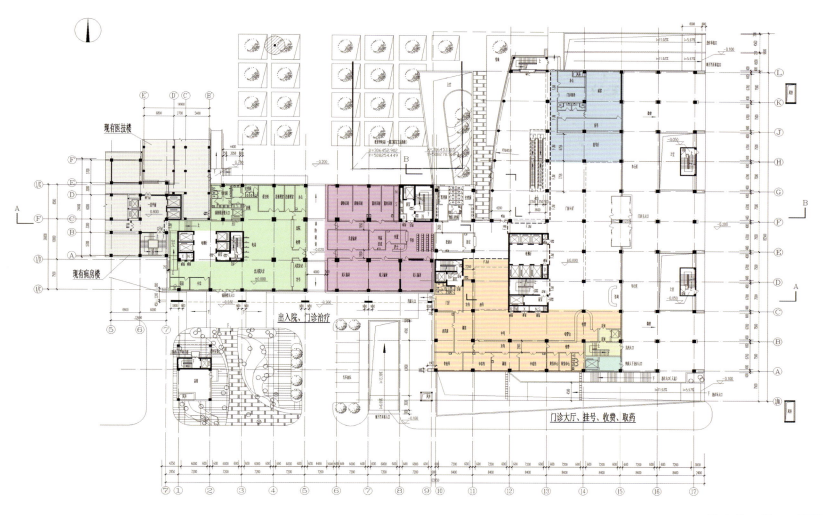

First Floor Plan 一层平面图

KEY WORDS 关键词

Concise Facade 立面简洁

Complete Function 功能齐全

Structured Layout 布局规整

Shenyang Air Force 1221 Project
沈空 1221 工程（文体医疗综合楼）

FEATURES 项目亮点

This is a project initiated and shared by Shenyang Air Force and the local government, and a fitness site to greet the 12th National Games. Functionally, it is divided into two fully functional parts: culture & sports and healthcare.

该项目是沈阳市军地共建共享的双拥项目，也是为了迎接第十二届全运会而建设的全民健身场所，功能上分为文体和医疗两个部分，各部分功能齐全。

Location: Shenyang, Liaoning, China
Architectural Design: China Northeast Architectural Design & Research Institute Co., Ltd.
Land Area: 8,963 m²
Total Floor Area: 27,438 m²
Plot Ratio: 2.9

项目地点：中国辽宁省沈阳市
建筑设计：中国建筑东北设计研究院有限公司
用地面积：8 963 m²
总建筑面积：27 438 m²
容积率：2.9

Overview

Located at Wanliutang Road, Liaohe District, Shenyang, this project is a nine-floor multiple-use building and boasts a total floor area of 27,438 m². It is a project initiated and shared by Shenyang Air Force and the local government, and a fitness site to greet the 12th National Games.

项目概况

项目为沈阳辽河区万柳塘路的一幢9层高的综合楼，总建筑面积为27 438 m²。该项目是沈阳市军地共建共享的双拥项目，也是为了迎接第十二届全运会而建设的全民健身场所。

Site Plan 总平面图

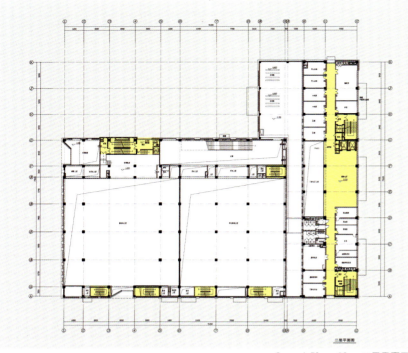
Second Floor Plan 二层平面图

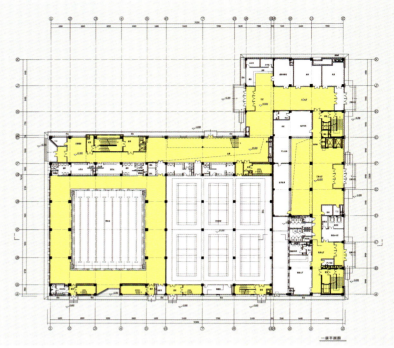
First Floor Plan 一层平面图

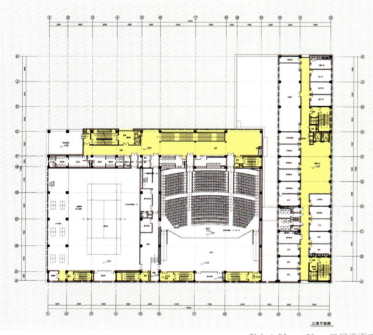
Third Floor Plan 三层平面图

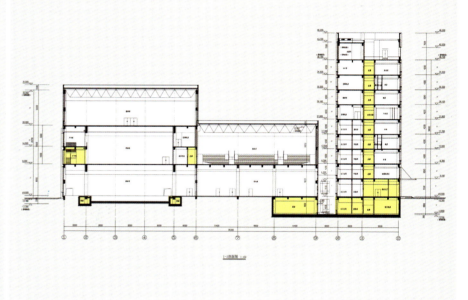
Sectional Drawing 剖面图

Planning

Functionally, this project is divided into two parts: culture & sports and healthcare. The side which borders Changqing Street is for pilot recruitment physical examination center and its entrance. It has nine floors overground and one floor underground. The first floor is for outpatient hall and emergency and the rest floors are for outpatient and examination center. The inner side is the culture & sports center, whose main body has three floors and part of it has two floors, and its main entrance is arranged on one side of Changqing Street. The first floor is for swimming pools and badminton hall, the second floor is for tennis court, table tennis hall and a conference hall which can accommodate 800 people, the third floor is for basketball hall and gym with a mezzanine. The underground floor is equipped with the equipment room, dining room, kitchen and some medical rooms.

规划布局

本项目功能上分为文体和医疗两个部分，其中临长青街一侧为招飞体检中心，入口设在临长青街一侧，地上九层、地下一层。其中一层为门诊大厅及急诊厅，其余各层为门诊及体检中心。靠近院内为文体中心，共三层，局部二层，主要入口设在长青街一侧，其中一层为游泳馆和羽毛球馆，二层为网球馆和乒乓球馆及一个能容纳800人的大会议厅，三层为篮球馆和健身馆，健身馆设有夹层。地下室为一层，设有设备用房、餐厅和厨房及部分医务用房。

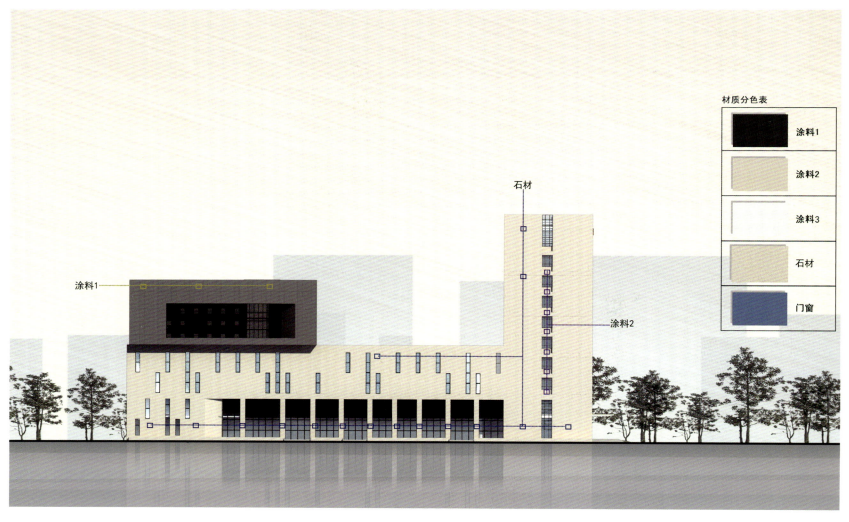

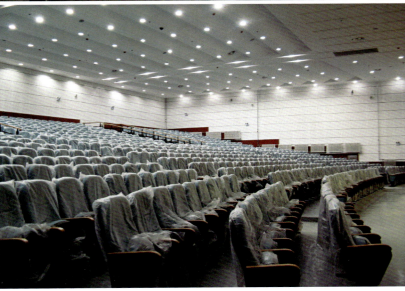

KEY WORDS 关键词

Quiet and Pleasant Environment 环境幽雅
Suitable for People 尺度宜人
Cellular Layout 单元式布局

Chongzhou People's Hospital and Chongzhou Maternal and Child Health Hospital

崇州市人民医院及崇州市妇幼保健院

FEATURES 项目亮点

In the elevation design, the whole hospital building group basically applies multilayer layouts to create quiet and pleasant environment. In the south side of the building group, the inpatient building of Chongzhou People's Hospital is heightened to be the landmark building for the whole district, connected with outpatient department but also relatively independent.

建筑形态上，整个医院建筑群基本采用多层布置的形式，营造宜人的尺度，给病人以亲切感。在建筑群南端，将人民医院的住院大楼加高，作为整个区域的统筹的标志性建筑，既与门诊联系又相对独立。

Location: Chongzhou, Sichuan, China
Architectural Design: Chongqin Architectural Institute
Total Land Area: 65,332 m^2
Total Floor Area: 57,186 m^2
The Floor Area of Chongzhou People Hospital: 45,187 m^2
The Floor Area of Chongzhou Maternal and Child Health Hospital: 11,999 m^2
Plot Ratio: 0.82
Greening Rate: 35.78%

项目地点：中国四川省崇州市
设计单位：重庆市设计院
项目总用地面积：65 332 m^2
项目总建筑面积：57 186 m^2
人民医院建筑面积：约 45 187 m^2
妇幼保健院建筑面积：约 11 999 m^2
容积率：0.82
绿地率：35.78%

Planning and Layout

The layout of the hospital building group focuses on the partitions for the infection areas and the activity and privacy areas. The western part is for the operation of Chongzhou People Hospital and the eastern part is for Chongzhou Maternal and Child Hospital. The main entrance of the hospitals is set in the north side of the area, close to the main urban passage, Yongkangdong Road. To completely split the flow of people and vehicles, these two hospitals set up the pedestrian plaza and the parking space separately so that the vehicles would not enter the pedestrian plaza. All the logistic lines and sewage transportation lines have been set with restricted land to enter the urban road without having to go far. Through fixed-time working system, the logistic lines and sewage transportation lines would not affect other hospitalized lines so as to avoid cross contamination.

规划布局

医院建筑群采用了多翼型+单元式布局，注重医院感染分区以及前后动静分区。西区为人民医院各部分功能楼，东区为妇幼保健院。医院主入口设置在场地的北面，靠近城市主干道永康东路。人民医院及妇幼保健院分别单独设置人行入口广场及停车场地，车行线路不进入人行广场，完全实现人车分流，医院所有物流和污物流线都考虑在建筑外侧设置专用通道，就近进入城市道路。通过定时作业，不与其他医疗流线冲突，避免了交叉感染。

Architectural Design

In the elevation design, the whole hospital building group basically applies multilayer layouts to create quiet and pleasant environment. In the south side of the building group, the inpatient building of Chongzhou People's Hospital is heightened to be the landmark building for the whole district, connected with outpatient department but also relatively independent.

建筑设计

建筑形态上，整个医院建筑群基本采用多层布置的形式，营造宜人的尺度，给病人以亲切感。在建筑群南端，将人民医院的住院大楼加高，作为整个区域的统筹的标志性建筑，既与门诊联系又相对独立。

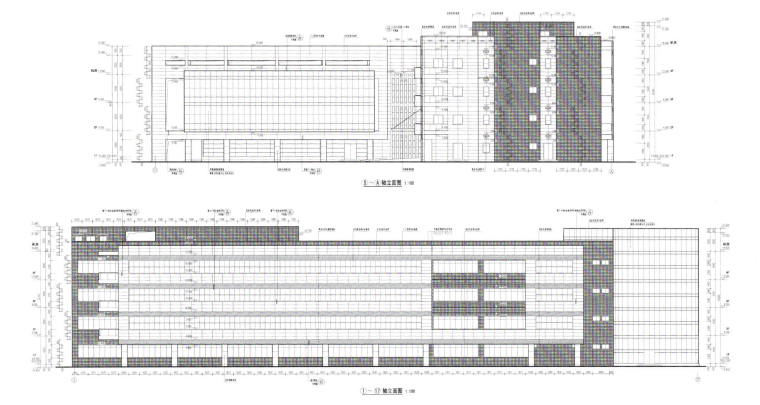

Side Elevation 轴立面图

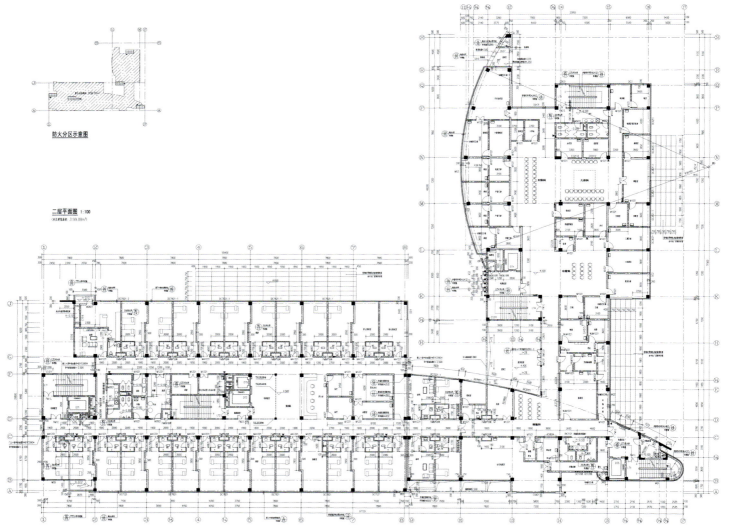

Second Floor Plan 二层平面图

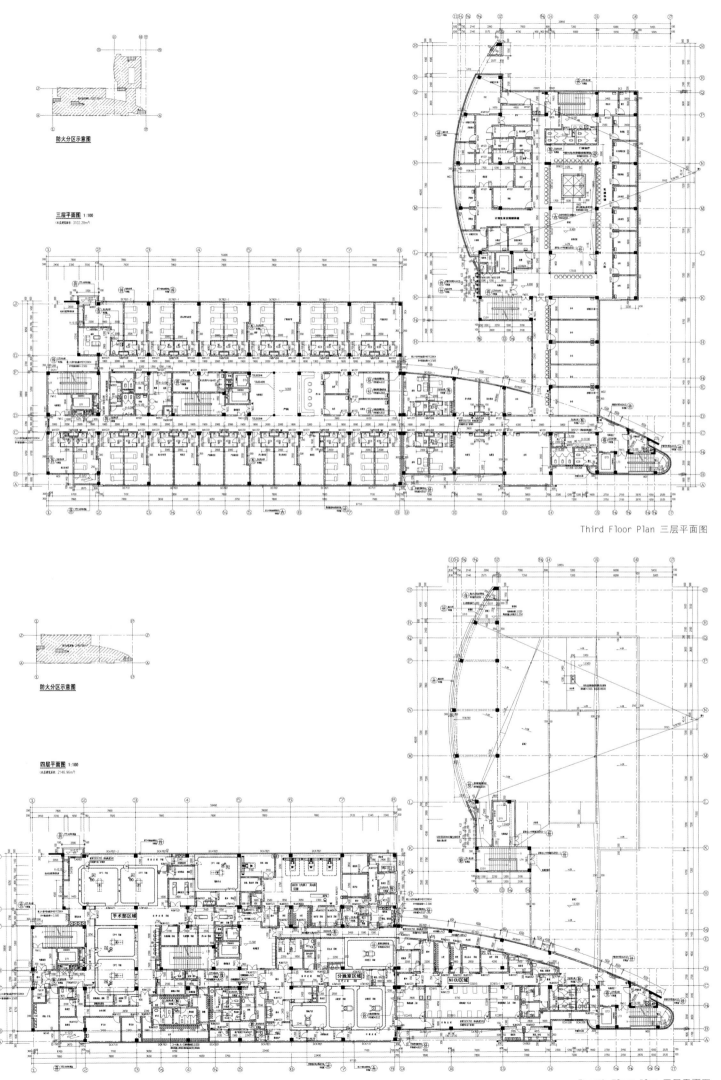

Third Floor Plan 三层平面图

Fourth Floor Plan 四层平面图

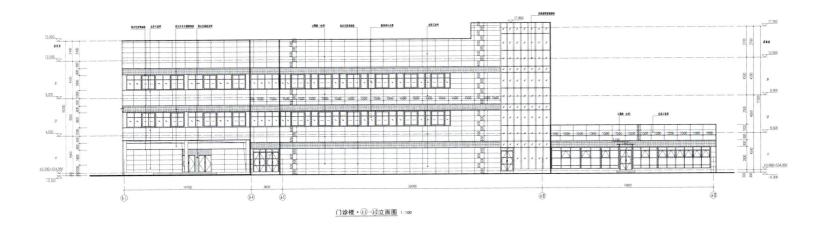

门诊楼·⑪-⑲立面图 1:100

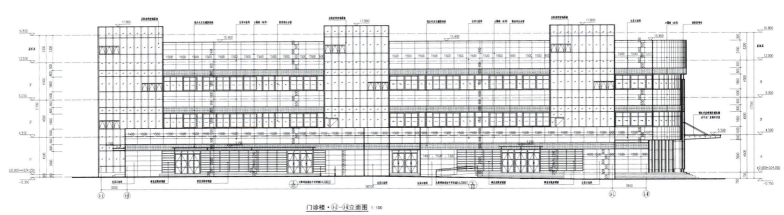

门诊楼·⑪-⑲立面图 1:100

Elevation 立面图

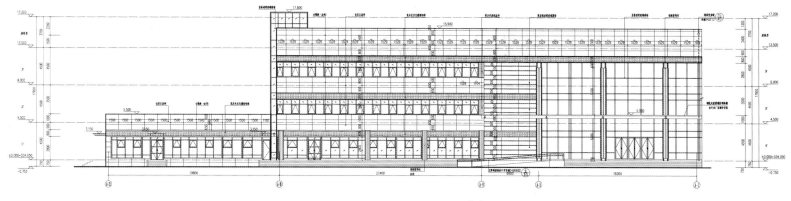

门诊楼·⑪-①立面图 1:100

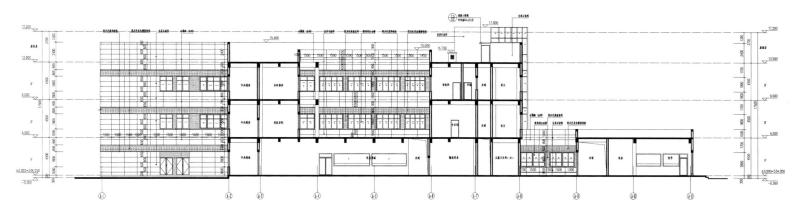

门诊楼·剖立面图 1:100

Elevation 立面图

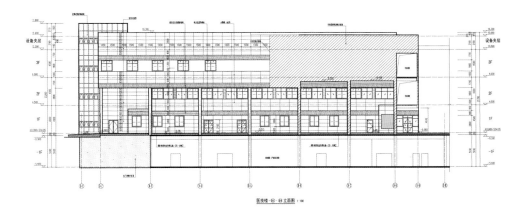

医技楼·⑨-⑥立面图 1:100

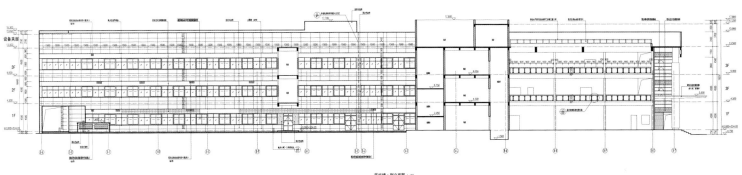

医技楼·剖立面图 1:100

Elevation 立面图

KEY WORDS 关键词

Modern Standard 现代标准
Clear Organization 分区明确
Harmonious and Diversified 和而不同

Outpatient Building of Jinan Military General Hospital
济南军区总医院门诊综合楼

FEATURES 项目亮点

The architects presented the proposal of "harmonious but different, coexisting with the surroundings, and keeping pace with the times", among which, "harmonious but different" means that the building should integrate into the surroundings and keep its own characteristics at the same time.

设计师们提出了"和而不同，与环境共存，与时代同步"的设计思路。"和而不同"，是指与周围现有建筑能和谐共处，但并非一味求同，以反映自身和时代特色。

Location: Jinan, Shandong, China
Architectural Design: Shangdong Tong Yuan Design Group Co.,Ltd.
Total Floor Area: 47,000 m²

项目地点：中国山东省济南市
设计单位：山东同圆设计集团有限公司
总建筑面积：47 000 m²

Overview

The new outpatient building of Jinan Military General Hospital is reconstructed on the original flat and open site, with the medical technology building on the north and the beautiful garden on the west. Surrounded by the ring road, the building will be the starting point of the hospital's main axis.

项目概况

济南军区总医院门诊综合楼是在原址拆除重建的工程，位于现医技楼南侧，西侧为环境优美的花园，周围是院区环形道路。整个场地平坦、开阔，建成后成为院区医疗主轴线的起点。

Overall Design Idea

Architectural design not only means the making of building form but also covers the creating of spaces and its surroundings. Especially for a re-constructed or extended project, it needs to inherit and innovate the "spirit of place". With this in mind, the architects presented the proposal of "harmonious but different, coexisting with the surroundings, and keeping pace with the times". "Harmonious but different" means that the building should integrate into the surroundings and keep its own characteristics at the same time.

总体设计思路

建筑设计并非是简单的形式问题，而是空间的营造，以及作为空间基础"场所"的营造。改扩建工程更需要延续并创新其特定环境的"场所精神"。为此，设计师们提出了"和而不同，与环境共存，与时代同步"的设计思路。"和而不同"，是指与周围现有建筑能和谐共处，但并非一味求同，以反映自身和时代特色。

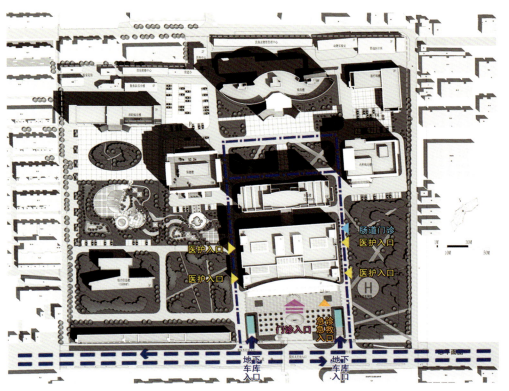

Traffic Network Analysis Drawing 交通流线分析图

Plan 平面图

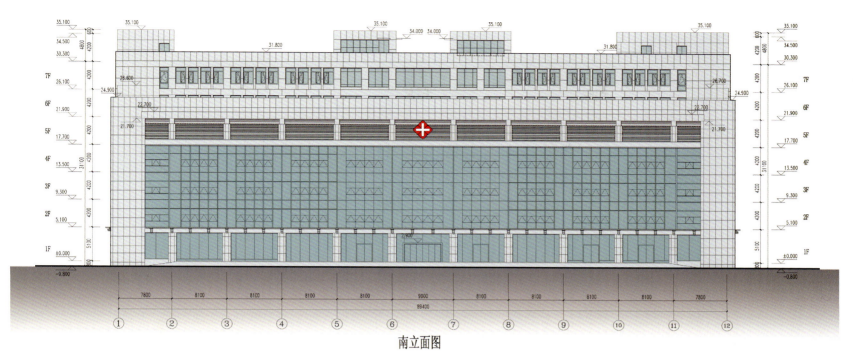

South Elevation 南立面图

West Elevation 西立面图

Traffic Flows

Because of the limitation of the plot size, the architects proposed to build an open square in front of the outpatient building. With entrances on two sides of the square, cars will be led to the underground parking directly from Shifan Road, thus the front square will be a safe space for pedestrians. The main entrance is barrier-free without steps for easy access. The entrances of the outpatient building and emergency department are set on the south respectively, staff-only entrances are on the east and west side, and the entrance for the intestinal is designed at the northeast corner.

交通流线

因用地前广场比较局促，本方案提出人车分流的开放式门诊广场概念。广场两侧的车库出入口直接从师范路进入地下车库，将整个院前广场打造成安全的人流广场。主出入口设计为无障碍入口，不设台阶，方便使用。门诊楼主入口，急诊和急救入口南立面依次展开；医护人员入口分别设于东、西两侧，东北角设肠道门诊入口。

KEY WORDS 关键词

- Green Building 绿色建筑
- Natural Environment 自然环境
- Lingnan Architectural Features 岭南建筑特色

Lingnan Hospital of the Third Hospital Affiliated Hospital, Sun Yat-Sen University (Luogang Central Hospital)

中山大学附属第三医院岭南医院（萝岗中心医院）

FEATURES 项目亮点

The project design pays attention to the local characteristics of Lingnan architecture, lays emphasis on the "people-oriented" design principle of construction details, so as to create garden-type building layout according to the characteristics of Lingnan architectural environment.

建筑设计中注重岭南建筑的地域特色，注重建筑细节的"以人为本"设计原则，营造出符合岭南建筑环境特点的花园式建筑布局。

Location: Guangzhou, Guangdong, China
Architectural Design: Guangzhou Urban Planning & Design Survey Research Institute
Total Land Area: 86,840 m²
Total Floor Area: 92,055 m²
Building Density: 19.6%
Plot Ratio: 0.88
Greening Ratio: 40%
Completion: 2011

项目地点：中国广东省广州市
设计单位：广州市城市规划勘测设计研究院
总用地面积：86 840 m²
总建筑面积：92 055 m²
建筑密度：19.6%
容积率：0.88
总绿地率：40%
竣工时间：2011 年

Overview

Lingnan Hospital of the Third Hospital Affiliated Hospital, Sun Yat-Sen University is a comprehensive AAA first-class hospital co-constructed by Luogang Government, Sun Yat-Sen University and the Third Hospital Affiliated Hospital of Sun Yat-Sen University. The hospital is located in Luogang New District with beautiful environment, green hills, green forest and fresh air.

项目概况

中山大学附属第三医院岭南医院是萝岗区政府、中山大学、中山大学附属第三医院合作共建的综合性三级甲等医院。医院位于环境优美的萝岗新区，院内青山环绕，绿树成林，空气清新。

Site Plan 总平面图

Planning and Design

The hospital planning fully respects and makes use of the natural environment, protects and takes good advantage of the original mountain to make the hospital building harmonious with the surrounding environment. The design adheres to the principle of sustainable development through methods of the land saving, energy saving, material saving and water saving, etc. to build a human-caring, comfortable and safe modern medical building.

规划设计

医院规划充分尊重和利用自然环境，竭力保护原有山体并加以利用，使医院建筑与周边环境共融。医院设计中坚持以可持续发展为原则，通过节地、节能、节材、节水等绿色建筑的设计手段，营造一个体现人性关怀、舒适、安全的现代化医疗建筑。

Architectural Design

The project design pays attention to the local characteristics of Lingnan architecture, lays emphasis on the "people-oriented" design principle of construction details, so as to create garden-type building layout according to the characteristics of Lingnan architectural environment.

建筑设计

建筑设计中注重岭南建筑的地域特色，注重建筑细节的"以人为本"设计原则，营造出符合岭南建筑环境特点的花园式建筑布局。

KEY WORDS 关键词

Hospital Street 医院街模式
Reasonable Organization 布局合理
Elegant Appearance 外观大气

Pengzhou People's Hospital
彭州市人民医院

FEATURES 项目亮点

The design is inspired by "the stem and branches of tree". Different buildings and facilities are connected together in this way with clear layout.

设计以"树干——树枝"的线状联系手法,把各门诊、急诊以及医技科室串联在一起,功能明晰,通过"医院街"的设计方便各种服务的定位,打破以往迷宫式的医院布局。

Location: Chengdu, Sichuan, China
Architectural Design: Fujian Provincial Institute of Architectural Design and Research
Building Height: 41.3 m
Total Land Area: 60,508 m²
Total Floor Area: 50,635.7 m²
Plot Ratio: 0.84
Greening Rate: 38.31%

项目地点:中国四川省成都市
设计单位:福建省建筑设计研究院
建筑高度:41.3 m
总用地面积:60 508 m²
总建筑面积:50 635.7 m²
容积率:0.84
绿化率:38.31%

Overview

The newly built Pengzhou People's Hospital, built by Xiamen Government, is located in Pengzhou South New Area. The buildings of the hospital are quite distinctive. The TCM building and outpatient building are linked by the "hospital street" to provide sufficient daylight and natural wind, and shorten the distance and pipelines between different departments. This kind of design has greatly saved the land and benefited the future development.

项目概况

新建的彭州市人民医院由福建厦门市援建,位于彭州市南部新区。医院的建设非常具有特色,中医技楼和门诊楼用"医院街"串联起来,在满足通风采光的前提下,有效缩短病人和医护人员的路程,同时缩短院区工程管线,是节地型的一种设计,为将来的发展提供了可能。

Hospital Street

"Hospital Street" reorganizes the functions and connects different buildings together to provide great convenience for staff and patients. In addition with the green belts along the hospital street, sunshine and green lands will accompany people. A garden-like hospital is created.

医院街模式

"医院街"是对功能的重新梳理,对流线的明晰组织。"街"把几个不同的形体串联在一起,为频繁来往于各功能用房之间的病人、医护人员提供了高质量的空间。同时,伴随着医院街做带状绿化,让阳光与绿地伴随病人、医护人员的每个行进过程,实现真正意义上的花园式医院。

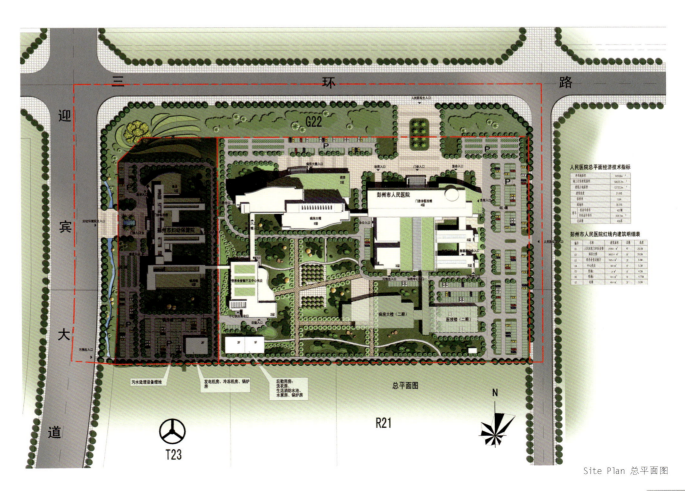

Site Plan 总平面图

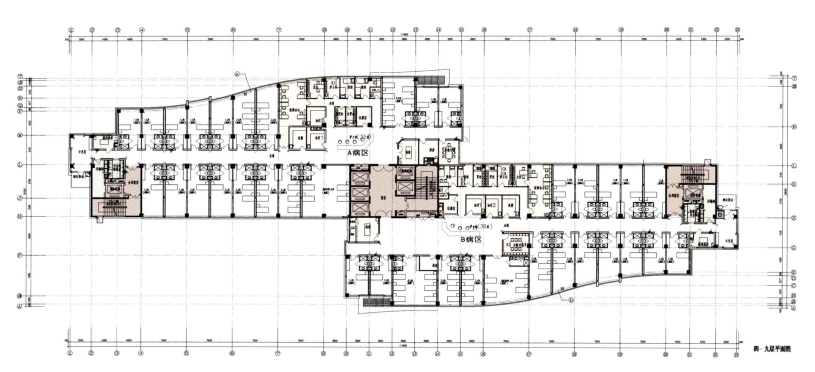

Ninth Floor Plan 九层平面图

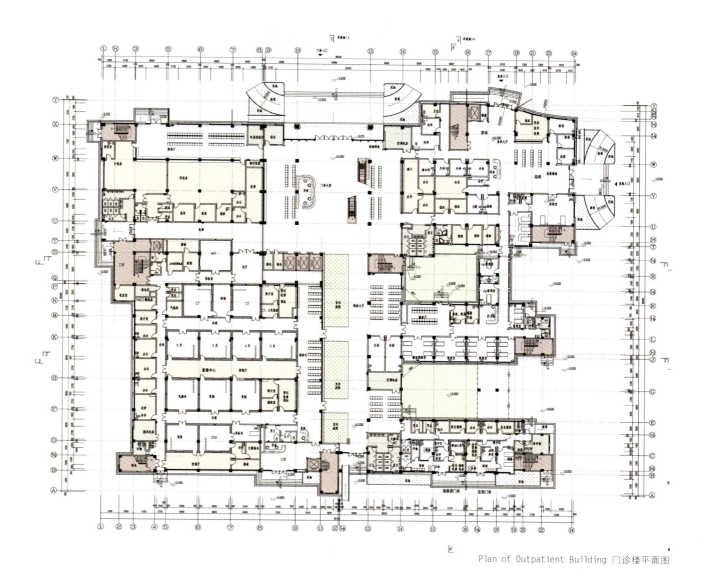

Plan of Outpatient Building 门诊楼平面图

Function Organization

The design is inspired by "the stem and branches of tree". Different buildings and facilities are connected together in this way with clear layout. At the same time, it well considers the activities of the patients and staff, and provides an independent access for the latter. In addition, the delivery of the clean items and sewages are clearly separated to ensure health and safety.

功能布局

设计以"树干——树枝"的线状联系手法,把各门诊、急诊以及医技科室串联在一起,功能明晰,通过"医院街"的设计方便各种服务的定位,打破以往迷宫式的医院布局。同时充分考虑病人、医护人员清晰的流线设计,为医护人员提供单独的通道。洁污物流做到明晰分流。

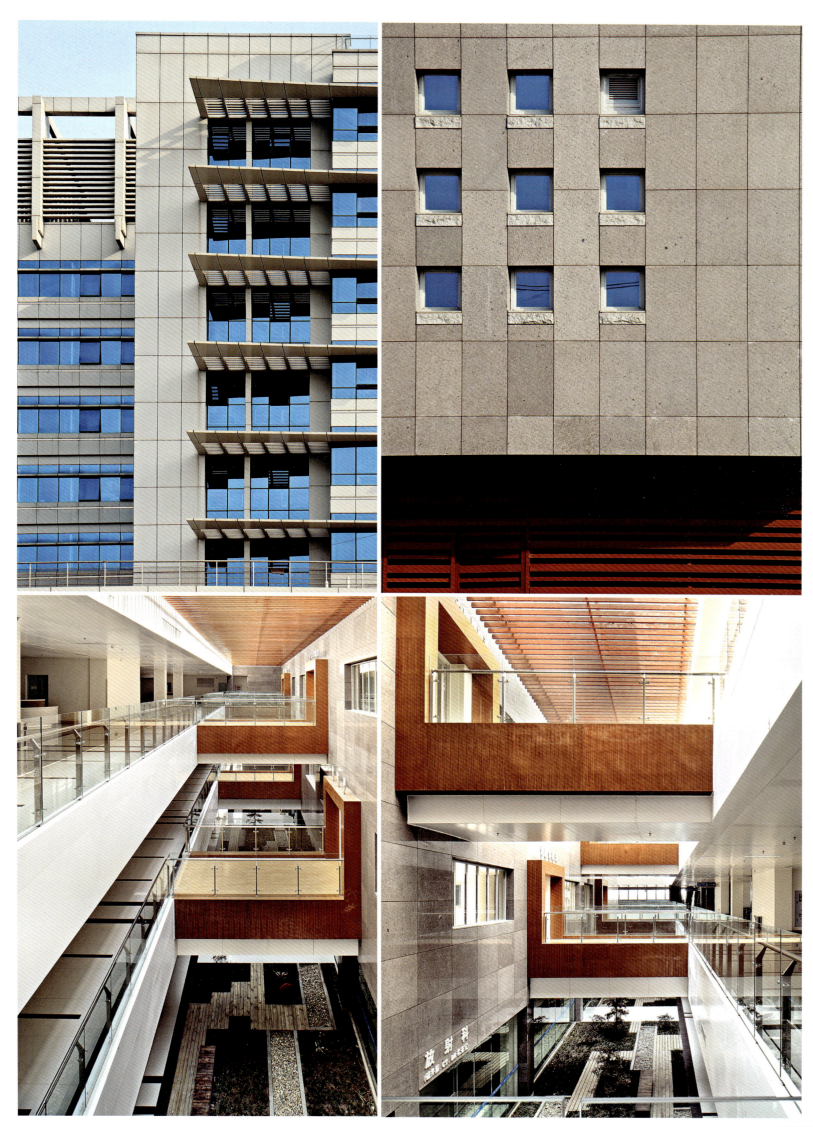